Fragile Computing

Laura Kocksch

Fragile Computing

How to Live With Insecure Technologies

palgrave
macmillan

Laura Kocksch
Department of Learning and Culture
Aalborg University
Copenhagen, Denmark

The book was originally defended as PhD thesis at the Faculty of Social Sciences, Ruhr University Bochum in June 2022.

ISBN 978-981-99-9806-7 ISBN 978-981-99-9807-4 (eBook)
https://doi.org/10.1007/978-981-99-9807-4

This Palgrave Macmillan imprint is published by the registered company Springer Nature Singapore Pte Ltd.
The registered company address is: 152 Beach Road, #21-01/04 Gateway East, Singapore 189721, Singapore

Paper in this product is recyclable.

To my family.

Acknowledgments

This research was funded by the NRW-Forschungskolleg "SecHuman—Security for Humans in Cyberspace" and the federal state of Nordrhein-Westfalen (Germany). The writing of this book was supported by Aalborg University (Denmark). No book is written alone, nor has academia ever been a lonely place for me. I am indebted to the participants of this study, to my PhD committee, and to many, many colleagues and friends: *Estrid Sørensen, Angela Martina Sasse, Torben Elgaard Jensen, Petra Ilyes, Gisela Welz, Matt Spencer, Lizzie Coles-Kemp, Mace Ojala, Alexander Helm, Katharina Kinder-Kurlanda, Alexander May, Andreas Poller, Mary Shnayien, Dominik Horrion, Georgi Iliev, Kerstin Hüttermann, the RUSTlab, and the TANTlab.*

Competing Interests No financial or personal conflicts of interests are to declare.

Ethics Approval All participants were informed of their participation in the study and signed participation agreements. All participants and institutions are pseudonymized. The study was not reviewed by an institutional ethics board as is common in German academia. The book underwent peer-review.

Contents

About the Author

Laura Kocksch defended her doctorate with honors ("summa cum laude") at the Ruhr University Bochum in June 2022. Since August 2022, she works as a post-doctoral researcher at the Techno-Anthropology Lab (TANTlab), Aalborg University, Copenhagen.

Kocksch completed her education in Cultural Anthropology and European Ethnology, Political Science and Sociology at the Goethe University in Frankfurt in 2016. Since 2014, she has been an active publishing author, reviewer, and publication chair for conference proceedings and journals in Computer Science, STS, and Anthropology. She conducts participant observation, interviews, and mixed-method ethnographies in various fieldsites such as social media, software development, critical infrastructures, universities, and small- and medium-sized enterprises. Kocksch engages with concepts in feminist STS, environmental anthropology and actor-network-theory.

List of Figures

1

Introduction

Between 2007 and 2010, the "Stuxnet" worm infiltrated industrial machines employed in nuclear energy plants in Belarus and Iran, becoming declared the first-ever cyber weapon.[1] In 2021, the breach of the Colonial gas and oil pipeline caused week-long panic buys and price explosions in the Southern US.[2] Computational systems underlying our critical infrastructures no longer stay idly in the background. They have begun to cause "drama" (Star, 1999).

In 2016, I became intrigued by the use of data science in two large-scale critical infrastructure companies in Germany. When getting field access and talking to data scientists, all they could talk about were the opportunities they were hoping to generate with new algorithms and data sources. Being an insurance and financial company, Covy felt pressure to modernize based on data companies venturing into the insurance industry. Google, they argued, would soon know their customers better than they do, having access to their daily search patterns instead of self-reported health statements. Powino, a subsidiary of one of Europe's

[1] https://www.wired.com/2014/11/countdown-to-zero-day-stuxnet/. For a material semiotic analysis of Stuxnet, see Balzacq and Cavelty (2016).

[2] https://www.bloomberg.com/news/articles/2021-06-04/hackers-breached-colonial-pipeline-using-compromised-password

L. Kocksch, *Fragile Computing*, https://doi.org/10.1007/978-981-99-9807-4_1

largest energy companies, argued that new data and data analytics had become essential to coordinate the shift from steadily producing coal and nuclear power plants to decentralized and weather-dependent renewable energy production. The idea circulated that with a modernized data infrastructure, they could remain relevant to an energy market that may sooner or later pivot to locally contained energy sharing economies.

Being educated in anthropology and science and technology studies (STS), I could barely contain my excitement to follow such innovation processes through the work of data scientists. And I could not have been luckier, having entered the companies right at a time when data scientists began building the computational infrastructure that was designed to support new data, tools, and access. When I began interviewing data scientists and joined meetings in their "labs", one issue continued to come up: Data scientists felt their work was held back or burdened with data security and data protection regulations. "The department of no" became a common explanation why a certain data set or tool could not be explored any further. Data scientists were annoyed, or outright furious, that their projects kept being rejected or cancelled by data security officers.

Luckily, they thought, they now had a scientist lurking around, and having no discernible qualifications to help with the data science, maybe she could be of some help with the pesky security department.

I had studied security requirements in developing practices before, assuring them that I had some sympathy for the difficulties to implement security department's wishes in software development (Poller et al., 2017). But my own interest was also piqued: Had I become pulled into the spirit of innovation and renewal that the labs were proclaiming? Within a few days, I had assumed that the most interesting story to tell about these companies was their data innovation, forgetting about the various efforts of developing technologies securely. I had become a manager, in the words of the post-actor-network-theory: I saw the companies through the keyhole of data science innovation—where technologies were planned anew, as if on a plain field; not through the eyes of the security departments where the tedious work of maintaining and repairing prevailed.

What started after this was a one-year exploration of the digital security practices of data scientists and security officers in the two companies. It is a timely moment to study digital security: Technologies, especially those being introduced newly, are subject to scrutiny of various kinds. Technologies are brittle and frail from their beginning, and continue to be patched up, updated, and prolonged.

Digital security is commonly portrayed as either a matter of technological finesse (writing "good" code, updating regularly, encrypting correctly), individuals' awareness (knowing about insecurity, and being incentivized to do something about it), or regulation (establishing rules). In practice, however, digital security is a site of collective negotiation, material action, delegation of responsibility, and establishing improvised, temporary, and fragile solutions. In practice, digital security illustrates technologies as *fragile*—more "good enough" than robust, more awkwardly patched together than planned out, more dependent on local adaptations than determining them.

The book traces digital security within, in tension and through the process of innovation in the labs, offering insights into ubiquitous, but often mundane practices of securing. Approaching digital security as practice opens it up for ethnographic observation. Four practices are portrayed over four chapters: testing (Chap. 2), tinkering (Chap. 3), training (Chap. 4), and performing (Chap. 5). The four practices are not encompassing, nor do they aim to be representative in any way. But they are not unique to the companies of this study either. Portraying one practice per chapter allows to stay with a way of securing for a while, without falling into the fallacy of judging if security is being done "right" here or there. The four chapters are not to be read independent, nor do they fully add up. They allow to discuss how securing practices conflate, conflict and are coordinated. They may stand in friction, sometimes partially connect. The goal is to "open up" digital security (Liebetrau & Christensen, 2021) by locating it in concrete material practices. By doing so, the chapters "populate" the matter of digital security with a plethora of objects, human- and non-human actors, action and inaction, as Bruno Latour would put it (Latour, 1992).

In contrast to the alluring world of data innovation that the labs so convincingly lured me into, the work of securing is slow, tedious, and "boring" (Star, 1999). Turning away from the innovative technology to mundane worlds of securing resonates with contemporary streams in STS and anthropology. By suggesting to engaging in fragile computing (rather than innovation computing, let alone computational innovation), we encounter technological practice differently.

Fragile computing does not mean "bad" computing: it does not mean practice that produces insecure technologies. Fragile computing encompasses practices aimed at attending to, maintaining and becoming affected by digital insecurity. Fragility thereby stands in contrast to "vulnerabilities" or "weaknesses"—more common terms in digital security research. What is fragile is precious and delicate; its frailness does not make it worth less, but often more. What is fragile is cared for not to eradicate weaknesses but to curate finesse.

Engaging with digital security as fragile computing allows to engage in competing goods involved in maintaining digital technologies. Securing entails continued worries, being affected by and responsive to technologies rather than solving contained security issues. The material work of securing is more ambiguous and conflicting, or "messier" (Jackson, 2014) than the world of innovation. Fragile computing is located in the realm of maintenance where what is locally "good" may appear "bad" from the distance, and vice versa (Danyi, 2022).

Perceiving of digital security practices as fragile computing allows focusing on mundane practice, their material components and often ambiguous morality. Where technologies are secured, things only partially stick, are improvised, contested, and blamed. Digital securities' "multiple ontologies", borrowing from Annemarie Mol (2003), spawn at such sites of practice. The resulting multiplicity can be particularly disconcerting in security matters: Do we not wish for a "baseline" of security to rely on? Of course! Multiplicity becomes uneasy, precarious, dangerous. This does not mean that we can dream it away. It is important to understand different enactments, not in order to commensurate them, but in order to cherish what is contested; understand its finesse and precarity.

The book concludes with a concretization of the term "fragile computing" by drawing together the four practices and literature in STS and anthropology. Dealing with fragility means noticing the various material flows, attending to what is frail but precious, fluid but refractory. Such attending is partially subverting, relational but not resolving, forges relations over time. Fragile computing does not call for generalist solutions but thinking that engages in ambiguities and dilemmas. How to live with insecure technologies includes living with fragility, crafting fragile relations and honing abilities to endure, be patient, do and undo. To STS readers, enduring fragility also entails conceptual challenges. It is uneasy to live in fragility which is why the book offers four tactics for research (Chap. 6).

STS and most prominently Infrastructure Studies have pointed out that repair and maintenance remain invisible to the users of infrastructures, this invisibility makes infrastructures seem frictionless and stable (Denis & Pontille, 2015). Being familiarized with every detail of operation may be hindering to the users of a system. On the other hand, it has been argued that any act of adapting (to) infrastructures may be perceived as an act of repair or maintenance in its own right; re-configurating the purpose and needs of a technology in perennial acts of use. Instead of understanding technologies as objects *on* which users act, it has been suggested to understand how they are being acted with, taking their material agency more seriously. To others, thinking with repair not only means decentering the active design from creators to maintenance workers and users but also involves a shift in valuing the entire lifespan of technology and not just their innovation and renewal. This line of thought comes particularly from information studies scholars who have observed a tendency among their own to focus on design, innovation, and deployment, but rarely on decommission, decay, or recycling (Cohn, 2016). In this growing landscape of scholars investigating the role of maintenance practices as well as end-of-life of technologies, recent anthropological literature finds a resonance. Anna Tsing (2015) introduces the notion of "ruins" to capture landscapes and more-than-human entanglements beyond repair and aims to train an "art of noticing" relations that unfold when restoration is no longer an option.

The notion of fragility is employed to point to two connected inquiries: On the one hand, it shifts attention to the instability of information technology—in contrast to perceiving information technologies as instruments of stability (or immutability, in the words of ANT). Whereas often designed to achieve universal functionality, interoperability, and compatibility, technologies are subject to situated action (Suchman, 2006), articulation work (Star & Strauss, 1999), and re-appropriation

(De Laet & Mol, 2000). While their stable configuration—allowing transmission of data, interoperability of programs, etc.—is imperative, they are also "fluid" objects of multimodal relating (De Laet & Mol, 2000). Their success may not lie in them being stable agents of normalization, but in allowing adaptation (Mol & Law, 1994), or their work as stable tools comes into being through them being mendable to others, for example, repair workers (Denis & Pontille, 2015).

The second line through which fragility emerges in is that in digital security debates, fluid forms become problematized and morally charged. When turning into a security risks, adaptations are no longer welcome, they pose a danger. The "human factor" is often considered the "weakest link" in digital security debates (Adams & Sasse, 1999; Squires & Shade, 2015). Re-articulation, re-purposing, and situated action are considered a source of insecurity. When employing the notion of fragile in this book, my intention is not to expose adaptation as weakness, but rather to describe its virtues. What is fragile is not what is ready-to-be-fixed. The practices described in this book do not aim at rectifying insecure information technology, nor are they a source of fragility that must be fixed. Fragile computing includes developing ways of becoming watchful and involved with technologies. There lies a tension here, I want to unfold a bit: If what I said was useful, and fragility emphasizes where adaptations and articulation of information technology is no longer considered playful adaptation of technologies, but judged as problematic, it suggests that it had been unproblematic so far. This is, of course, only partially true as studies in post-ANT have argued. Being not the intended user of a system has always been a source of tension (Star, 1990), but now these tensions become more pertinent and shameful.

Another aspect of the tension is that in infrastructures, maintenance work is considered invisible and staying in the background, but in fragile computing, they become vocal and demanding. So far so that they begin to enroll users—those that were in the privilege of "just" enjoying the stability of the systems when they did not engage in repair work (Denis & Pontille, 2015). This unfurls the topologies of information technologies introduced by ANT and post-ANT further: immutable stability becomes trenched with demands of maintenance and mutable fluidity becomes risky. It is, thus, necessary to take fragility seriously: its reach and morality.

Part of this endeavor is to halt judgment for a moment. For example, judgment of users' articulation work. When divergence of designed purposes become a security risk, scholarship in actor-network-theory (ANT) suggests asking whether we should aim for utopian strength or try to understand more fluid forms (Law, 2000; Law & Mol, 2002a, 2002b). In such, security oscillates, is ambiguous and complicated. Halting judgment means not to avoid this complication, but to engage with it actively. I put forward an understanding of such fluid and uneasy digital security; or where computational practice becomes *fragile*.

The empirical chapters unfold how security is negotiated, partially achieved, postponed, distributed, etc. while staying responsive to the need to secure. We can neither overlook such situated fluid action, nor can we overlook security requirements. But commonly, security is alien to a fluid thinking: security requires discerning between areas of security and insecurity, building "parameters" (the military narration is obvious), or more recently "zero-trust"-relations (Spencer & Pizio, 2023), which, in turn, strengthens the imagination of technologies as being located "in" the social world (a company, a school, a car) with clear boundaries between the two (Sørensen, 2009). Topologies that are imperative to securing may be partially incommensurable to fluid forms of technological adoption as put forward in ANT. A sensitivity to this trouble of topologies is needed. Fragility includes this trouble: What is fragile is both fluid and contested, both precious and precarious.

Fragility has been discussed in relation to the temporality of infrastructure, for example, their decay over time (Ramakrishnan et al., 2021), or active tempering (Denis & Pontille, 2015). As a material condition, fragility points to the agency of objects, and trains abilities to recognize and attend to materiality (Denis & Pontille, 2023). This requires a shift in thinking from innovation and renewal toward repair and maintenance (Jackson, 2014). Fragility is not only a material condition but also a way of thinking about living together with decaying things and infrastructures. To Mol (2008), accepting fragility is part of the "logic of care" that is distinct from the logic of choice. In the logic of choice, informed rational subjects make decisions, separating them from the complications of the issue of hand. In the logic of care, it must be endured that not all things are rectifiable. Living with fragility then means engaging with mess—both in the sense of maintenance work (where things get dirty), and in the sense of moral ambiguity (where things are not simply right or wrong). While as material condition, fragility can be the outcome of isolated incidents (vandalism or accidents), or gradually (over time), in the reading put forward here, it is also "normal", but uneasy to admit to (Perrow, 1999).

The book offers four situated enactments of digital security. In two, digital security is marginal, and secondary, epitomizing it as invisible and subversive work, while in the others, it is put on stage, rendered explicit in performances of testing and training. While each chapter presents socio-material practices through which digital security is enacted, taken together, they provide an impression of the multiplicity of the object of study and the need for not one but a variety of conceptual approaches to open digital security for studies in STS.

- **Testing** demonstrates how digital security is performed in different shapes or "patterns of relation" (Sørensen, 2009, p. 70). Applying such a topological lens inspired by STS literature allows us to specify how practitioners perform digital security in different forms, each producing a different outcome of what "good" security means and can be assessed. Testing practices do not solely aim at producing assurance but must also endure ambiguity and inconclusiveness.
- Following two developers while they write access control policy code, **Tinkering** offers an opportunity to learn how digital security is enacted as meticulous and situated practice that is engaged with conflicting moralities. Setting access control policy code is a material and moral practice. It requires practices of trying and failing, honing one's abilities of noticing, overflowing formal organizational roles and creating long-term relations. The chapter proposes to perceive of such handling of fragility as involving care. Caring is not aimed at solving the difficulties of setting access code once and for all, but about negotiating what is acceptable in a specific place and time. Care—drawing on the work of doctors and nurses—involves bending the rules and turning advice into livable or workable guidance for specific patients (in this case colleagues). Caring means handling fragility, where there is no light at the end of the tunnel.

- Reflecting upon my participation in an emergency **Training**, digital security in this chapter is discussed as both an event, and a long-term organizational practice. Before the training, many efforts go into creating a digital security emergency, for example, secrecy, warning texts, and measuring response times. During the training, participants, however, rely on knowledge they have built long-term and in attuning to each other. During the training, digital security becomes a matter of testing out options against previously acquired knowledge and attempts to limit harm. Instead of prevention, the training is aimed at preparedness for an emergency that reaches beyond its time-limited occurrence. In the ceremonial exercise, digital security includes creating an organizational stand-by—a state of attention to what is not to prevent.
- Finally, by looking at how digital security contributes to changing practices of **Performing**, the last chapter points to how digital security relates to practices of data innovation. The chapter follows the transition of the companies from a domain-specific to a centralized data storage and management, laying the foundation for new data science applications. In processes of innovation, digital security has an ambivalent function; both creating resilience and warning, supporting renewal while reminding to slow down. The chapter unfolds how digital security is pivotal in building trust in numbers and developing relevant inquiries for other's data.

By focusing on the four practices, the book offers accounts of enacting digital security—performing it locally, and often diverging from standard definitions. Each practice negotiates what security means in a particular setting, sometimes more obviously (by setting numeric standards of secure/insecure), sometimes less obvious (by trialing and testing until a security function is "good" enough).

Each chapter outlines different practices of doing digital security—some of which may lie in the realm of security experts, others in the area of other fields. I have made no difference between the two as my aim was to focus on the concrete material practices of enacting digital security, not on a judgment of whether such security was done "right" (according to regulation). What is fragile is prone to break down, it may not persist.

Developers, security officers, and data scientists deal with this possibility of breakdown and ending. This is not a nice and cozy world, you may think. It is a world where things go wrong, where people are blamed. But it is the world where digital security is configured in action.

The episodic approach taken here assumes that there is not one whole of digital security to be captured, and objective reality to elicit, but rather that "it" is enacted through situated practices. In "enactments"—the practicalities in which ontologies are bound up—Mol (2003) suggests, "there is no longer a single passive object in the middle, waiting to be seen from the point of view of seemingly endless series of perspectives. Instead, objects come into being—and disappear—with practices" (p. 5). The four practices portray digital security not as a matter of incompliance from a pre-set definition of security, but more precisely as sites where multiple securities emerge, meet and conflict.

Ontologies of digital security become "coordinated", conflate and may contradict (Mol, 2003). Digital security becomes *multiple*, yet not plural. Connections between the chapters are noted, yet many frictions remain (Tsing, 2004). The chapters are isolated but also connected—through the companies they are part of or through their common interests in securing. But the chapters are also in isolation and can be partially contradictory or antagonistic. Not all of them come together in a coherent way. To Mol, such valuing of practices demands the renaming ethnography into "praxiography" (Mol, 2003; Heuts & Mol, 2013).

A methodological question ensues: How should one describe digital security as fluid practice that is observable to praxiographic methods? Methodologically, this requires getting access to and staying with technological practice for such an ephemeral object like security to "appear" (and "disappear") or for interlocutors to point it out to us (or point out its absences). It requires attention to what is hidden away and rendered organizationally invisible. And it asks for the building of rapport and noticing of awkwardness and moral imperatives induced by the presence of a researcher who investigates security. After all, making visible in/security is risky; an ethnography of it could be itself "hacking" that exposes insecurities. Ethnography means recording, taking notes, and documenting what is often unspoken (Hirschauer, 2001). It requires gaining access to practices that are tediously concealed in order to be effective (Sundaramurthy et al., 2014).

To become part of a company, one has to follow its security guidelines, or put differently, following digital security guidelines makes you become part of the company. Throughout this journey (that requires deference to security rules), an ethnographer is able to trace why and for what reason workarounds or contentions are performed, which, in turn, improves our understanding of incompliance and "problematic" behavior. Ethnographic details can complement previously mentioned approaches because it emphasizes the situated and continuous efforts and troubles. Digital security lies in the everyday and is often approached in between or in addition to other things. These can only be observed when being present in person and feeling and going along with people's life and work.

My presence in the field, informants, scientific colleagues, and materials all co-created the research as I present it here. Taking the contingency of my ethnography as a starting point helps to describe my ethnographic journey, challenges, and moments of surprise. Paying attention to the practicalities of the ethnography endorses the situatedness of the researcher, where she comes from, what she intends, and also how intentions shift. Ethnographers are often confronted with complicated decisions and practical constraints that are rarely captured in Method handbooks. Attending to the "here-and-now practicalities" (Sørensen, 2009, p. 10) is important to present the ethnographic method as contingent and not simply resting on the ethnographer's choices. In my case, these practicalities included wrestling with my object of study occasionally: Gaining access as ethnographer while not becoming a hacker or "spy" (Boas, 1919) myself. While "co-laborative" research in STS and anthropology has been praised for moving ethnographic practice away from the armchair and into the direct collaboration with those being studied, studying corporations is rarely considered (Niewöhner, 2016). The insurance and energy industry, in particular, may be rather critiqued upon than allied up with by anthropologists or STS scholars. A "mutual skepticism" (Liburkina, 2021) may even exist. As the description of entering the field earlier in this chapter has proposed, however, I did become a collaborator to the field: they hoped I could mediate in a conflict between data scientists and security departments. And I attempted to live up to their expectations by presenting results of my study in both arenas, aiming to familiarize each with the other's work practices and incorporated logics (e.g., of playfulness and experimentation in data science and of assurance and maintenance in digital security). The embodied presence of the ethnographer as a third party in this setting allowed for modes of co-laboration with industry actors that may not align with my own views or agendas. While collaborations with activists and even policy makers in STS and anthropology are more common (as political goals more often align), the willingness for situated co-laboration may be required to study fieldsites like mine, and I did not feel betraying my beliefs in moderating a looming friction in the field.

A second challenge emerged, however. While ethnographies of digital security are still scarce (Dourish et al., 2004; Squires & Shade, 2015; Kocksch et al., 2018), they have proven valuable to investigate individual's mental models (Sundaramurthy et al., 2014), group's opinions and perceptions of digital security (Palombo et al., 2020), or organizational responses to changed security policies (Poller et al., 2017). However, following debates in Human-Computer Interaction, anthropological ethnography is not solely a "tool" that can be added to existing methods of extracting information from a field (such as interviews or quantitative surveys) (Dourish, 2006).[3] Rather, ethnography has a commitment to develop "theoretical descriptions" (Hammersley, 1990): it aims to conceptualize through empirical stories, instead of fitting into theories off the shelf. Put differently, ethnographic engagement with digital security means unsettling its established preconceptions, and ontology. Instead of "applying" ethnography to pre-established understandings of digital

[3] Computer scientist Paul Dourish insists that computer science should take ethnography's conceptual contribution more seriously, for example, that there is not one "truth" to uncover about technological practices, that they are enacted in multiple ways (Dourish, 2004). Ethnographers are not "tape recorders" to bring the "reality" of technological use into engineering labs. Part of the debate is whether engineers can "simply" acquire the competencies of conducting the material work of ethnography without becoming familiar with the origins of ethnography in critical anthropological research (Forsythe, 1999). Reducing ethnography to a set of techniques may "underestimate, misstate, or misconstrue the goals and mechanisms of ethnographic investigation" (Dourish, 2006, 542). To Dourish, assessing ethnography by how well it develops "implications for design" falls into a fallacy as it underestimates ethnographer's abilities to uncover underlying assumptions and commitment that things could be different. As anthropologist Marilyn Strathern writes: We must understand ethnography's role "not as adding more of the same [...] but as the intermeshing of different orders of phenomena, having to take certainties and uncertainties together" (Strathern, 2002, 312).

security, I understand ethnographic practice as a mode of theorizing digital security differently. This also means that it is not a sole method of data extraction from a field, but carefully engages with a field in moving along while also situatively juxtaposing oppositions of the field. It is a practice of both immersion into and alienation of the parties in a field (Emerson et al., 2011). This book is written as a series of practices that overflood the pre-set ontology of digital security. Instead of asking how well actors "perform" or "comply", these practices are "opening up" digital security (Liebetrau & Christensen, 2021).[4] What is secure/insecure becomes an empirical question. Other questions result from it: What is "good"/"bad" security in a specific practice? How to live with less-than-optimal security? and so on. The contribution of my ethnographic study of digital security does not lie in providing solutions, that is, by explaining human behavior, but in crafting stories that immerse us, and allow us to halt a moment, be surprised and even stumble.

Although it is imperative to understand that ethnography is not a tool that can be judged by how well it produces insights for existing digital security research, the book offers nine implications that should be both useful for readers in computer science and in STS or anthropology. By choosing the format of implications, the book aims to abstract from the practices described here and suggest that, although not aiming for representativeness, the empirical stories and the conceptual repertoires that are developed through them is relevant "elsewhere". Implications will have to be tested and trialed further.

Each of the empirical chapters puts the focus on one practice of digital security: testing, tinkering, training, and performing. Each empirical chapter develops terms and concepts in relation to the empirical material. This performs the work of theorizing digital security as practice, while staying partially inconclusive. Perceiving of concepts not as exclusive and

[4] Ethnographic writing, anthropologist Anand Pandian notes in reference to Kathleen Stewart, "tries to let the otherwise break through, to keep it alive, to tend it" (Pandian, 2019, 7). Ethnographies are exactly that: they tell different stories than other methods; they focus on what is not easily focused on. Or as Pandian puts it, they "rob the proud of their surety and amplifies voices otherwise inaudible" (ibid.). "Writing with care", he continues, is a form of "letting things be vulnerable and uncertain" (Pandian, 2019, 14).

convoluted, but as "companions" (Winthereik, 2020)[5] emphasizing some nuance (leaving aside others) creates a needed arena to theorize digital security. This can be understood as a playful engagement with theories, testing and trialing them in conversation with the empirical stories and interlocutors. This playfulness is intended: What does the "use" of concepts (for lack of better terminology) do to the field; to its relations and in/consistencies?

Conceptually, the four chapters offer (symmetrically) descriptions of practices that enact "expert" versions of digital security such as testing and trainings, as well as more subversive versions such as haggling over access control policies or negotiating data confidentiality.

I could have pointed to similar themes in each practice, but decided to give them more characteristic traits, while also provoking before-established terms and categories across them. Taken together, the chapters, however, do not convolute into a new theory of digital security. The goal is to open up digital security for STS and anthropology by offering some conceptual tools (others are possible and encouraged). This explores the additive value of thinking with tried and tested STS themes in connection with digital security. This does not attempt a critique per se on existing approaches in digital security (nor on STS concepts, for that matter) but aims to contribute new heuristics to the respective repertoires. This is not devaluation of existing approaches to digital security, such as usable security, mathematics, and engineering (in fact, those are imperative in many regards), but simply an opening of a new conversation about digital security based on the realization that the issue may be of increased complexity. The book equips us with examples, ethnographic anecdotes, and concepts to further our understanding of digital security in the current time. It argues for the purchase of multiple conceptual repertoires that are partially, but not fully, commensurable. Reiterating a commitment in STS and ethnography to opening seemingly settled definitions by rendering them as empirical objects, this book suggests new vocabulary and alternative modes of thinking and doing digital security.

[5] Brit Ross Winthereik suggests "concepts as companions" that travel with us through the field and become—like the ethnographer—changed through the encounter. "Concepts companions can help open worlds, but they can also be too loud and talkative to the already quite heavily populated places we visit during ethnographies" (Winthereik, 2020, 30). Concepts are not only companions to ethnographers but interlocutors as well. Ethnographers are sensitive to such concepts in the field, as well as their contradictions or hierarchies. This sensitivity allows for juxtaposition of concepts that are prevalent in the field (e.g., definitions of security) with concepts that we "bring".

References

Adams, A., & Sasse, M. A. (1999). Users are not the enemy. *Communications of the ACM, 42*(12), 40–46. https://doi.org/10.1145/322796.322806

Balzacq, T., & Cavelty, M. D. (2016). A theory of actor-network for cybersecurity. *European Journal of International Security, 1*(2), 176–198. https://doi.org/10.1017/eis.2016.8

Boas, F. (1919). Scientists as spies. *The Nation, 27*. https://doi.org/10.1111/j.0268-540X.2005.00359.x

Cohn, M. L. (2016). Convivial decay: Entangled lifetimes in a geriatric infrastructure. In *CSCW'16 Proceedings and companion of the ACM conference on computer-supported cooperative work and social computing*, San Francisco, CA, pp. 1511–1523.

Danyi, E. (2022). *Melancholy democracy: Politics beyond hope and despair.* Habilitation submitted, Institut für Soziologie, Goethe-Universität.

De Laet, M., & Mol, A. (2000). The Zimbabwe bush pump. *Social Studies of Science, 30*(2), 225–263. https://doi.org/10.1177/030631200030002002

Denis, J., & Pontille, D. (2015). Material ordering and the care of things. *Science, Technology, and Human Values, 40*(3), 338–367. https://doi.org/10.1177/0162243914553129

Denis, J., & Pontille, D. (2023). Cultivating attention to fragility: The sensible encounters of maintenance. In *Ecological reparation* (pp. 344–361). Bristol University Press.

Dourish, P. (2004). What we talk about when we talk about context. *Personal and Ubiquitous Computing, 8*(1), 19–30. https://doi.org/10.1007/s00779-003-0253-8

Dourish, P. (2006). Implications for design. In R. Grinter (Ed.), *Proceedings of the SIGCHI conference on human factors in computing systems* (pp. 541–550). ACM.

Dourish, P., Grinter, R. E., La Delgado Flor, J., & Joseph, M. (2004). Security in the wild: User strategies for managing security as an everyday, practical problem. *Personal and Ubiquitous Computing, 8*(6), 391–401. https://doi.org/10.1007/s00779-004-0308-5

Emerson, R. M., Fretz, R. I., & Shaw, L. L. (2011). *Writing ethnographic fieldnotes* (2nd ed.). The University of Chicago Press.

Forsythe, D. E. (1999). "It's just a matter of common sense": Ethnography as invisible work. *Computer Supported Cooperative Work, 8*(1–2), 127–145. https://doi.org/10.1023/A:1008692231284

Hammersley, M. (1990). What's wrong with ethnography? The myth of theoretical description. *Sociology, 24*(4), 597–615. https://doi.org/10.1177/0038038590024004003
Heuts, F., & Mol, A. (2013). What is a good tomato? A case of valuing in practice. *Valuation Studies, 1*(2), 125–146. https://doi.org/10.3384/vs.2001-5992.1312125
Hirschauer, S. (2001). Ethnografisches Schreiben und die Schweigsamkeit des Sozialen/Ethnographic Writing and the Silence of the Social. *Zeitschrift für Soziologie, 30*(6), 429–451. https://doi.org/10.1515/zfsoz-2001-0602
Jackson, S. J. (2014). 11 Rethinking repair. In Gillespie, T., Boczkowksi, P. & Foot, K. A. (Eds.) *Media technologies: Essays on communication, materiality, and society* (pp. 221–239). https://doi.org/10.7551/mitpress/9042.003.0015.
Kocksch, L., Korn, M., Poller, A., & Wagenknecht, S. (2018). Caring for IT Security: Accountabilities, moralities, and oscillations in IT security practices. *Proceedings of the ACM on Human-Computer Interaction, 2*(CSCW), Article 92. https://doi.org/10.1145/3274361.
Latour, B. (1992). Where are the missing masses? The sociology of a few mundane artifacts. In W. Bijker & J. Law (Eds.), *Shaping technology-building society: Studies in sociotechnical change* (pp. 225–259). MIT Press.
Law, J. (2000). *Ladbroke Grove, or how to think about failing systems*. http://www.comp.lancs.ac.uk/sociology/papers/Law-Ladbroke-Grove-Failing-Systems.pdf.
Law, J., & Mol, A. (2002a). Local entanglements or utopian moves: An inquiry into train accidents. *The Sociological Review, 50*(1_suppl), 82–105. https://doi.org/10.1111/j.1467-954x.2002.tb03580.x
Law, J., & Mol, A. (Eds.). (2002b). *Complexities*. Duke University Press.
Liburkina, R. (2021). Extraordinary ethnographic encounters in extraordinary times: A plea for experimental interventions in more-than-business relations. *Kulturanthropologie Notizen, 83*, 14–26.
Liebetrau, T., & Christensen, K. K. (2021). The ontological politics of cyber security: Emerging agencies, actors, sites and spaces. *European Journal of International Security, 6*(1), 25–43. https://doi.org/10.1017/eis.2020.10
Mol, A. (2003). *The body multiple*. Duke University Press.
Mol, A. (2008). *The logic of care*. Routledge.
Mol, A., & Law, J. (1994). Regions, Networks and Fluids: Anaemia and Social Topology. *Social Studies of Science, 24*(4), 641–671. https://doi.org/10.1177/030631279402400402

Niewöhner, J. (2016). Co-laborative anthropology: Crafting reflexivities experimentally. *Etnologinen tulkinta ja analyysi*, 81–124. (in Finnish).

Palombo, H., Ziaie Tabari, A., Lende, D., Ligatti, J., & Ou, X. (2020). An ethnographic understanding of software (in) security and a co-creation model to improve secure software development. In *Proceedings of the sixteenth USENIX conference on usable privacy and security (SOUPS'20)* (pp. 205–220). USENIX Association.

Pandian, A. (2019). *A possible anthropology: Methods for uneasy times*. Duke University Press.

Perrow, C. (1999). *Normal accidents*. Princeton University Press.

Poller, A., Kocksch, L., Türpe, S., Epp, F. A., & Kinder-Kurlanda, K. (2017). Can security become a routine? A study of organizational change in an agile software development group. In *Proceedings of the 2017 ACM conference on computer supported cooperative work and social computing (CSCW '17)*, 2489–2503. Association for Computing Machinery. https://doi.org/10.1145/2998181.2998191

Ramakrishnan, K., O'Reilly, K., & Budds, J. (2021). The temporal fragility of infrastructure: Theorizing decay, maintenance, and repair. *Environment and Planning E: Nature and Space, 4*(3), 674–695. https://doi.org/10.1177/2514848620979712

Sørensen, E. (2009). *The materiality of learning: Technology and knowledge in educational practice (Learning in doing: Social, cognitive and computational perspectives)*. Cambridge University Press. https://doi.org/10.1017/CBO9780511576362

Spencer, M., & Pizio, D. (2023). The de-perimeterisation of information security: The Jericho Forum, zero trust, and narrativity. *Social Studies of Science*, 3063127231221107. Advance online publication. https://doi.org/10.1177/03063127231221107

Squires, S., & Shade, M. (2015). People, the weak link in cyber-security: Can ethnography bridge the gap? *Ethnographic Praxis in Industry Conference Proceedings, 1*, 47–57. https://doi.org/10.1111/1559-8918.2015.01039

Star, S. L. (1990). Power, technology and the phenomenology of conventions: On being allergic to onions. *The Sociological Review, 38*(1_suppl), 26–56. https://doi.org/10.1111/j.1467-954X.1990.tb03347.x

Star, S. L. (1999). The ethnography of infrastructure. *American Behavioral Scientist, 43*(3), 377–391. https://doi.org/10.1177/00027649921955326

Star, S. L., & Strauss, A. (1999). Layers of silence, arenas of voice: The ecology of visible and invisible work. *Computer Supported Cooperative Work (CSCW), 8*, 9–30. https://doi.org/10.1023/A:1008651105359

Strathern, M. (2002). Abstraction and decontextualization. An anthropological comment. In S. Woolgar (Ed.), *Virtual society?: Technology, cyberbole, reality* (pp. 302–314).

Suchman, L. (2006). Human-machine reconfigurations: Plans and situated actions. In *Learning in doing: Social, cognitive and computational perspectives* (2nd ed.). Cambridge University Press. https://doi.org/10.1017/CBO9780511808418

Sundaramurthy, S. C., Case, J., Truong, T., Zomlot, L., & Hoffmann, M. (2014). A tale of three security operation centers. In *Proceedings of the 2014 ACM workshop on security information workers (SIW' 14)* (pp. 43–50). Association for Computing Machinery. https://doi.org/10.1145/2663887.2663904

Tsing, A. L. (2004). *Friction*. Princeton University Press.

Tsing, A. L. (2015). *The mushroom at the end of the world: On the possibility of life in capitalist ruins*. Princeton University Press.

Winthereik, B. R. (2020). Is ANT's radical empiricism ethnographic? In A. Blok, I. Farias, & C. Roberts (Eds.), *The Routledge companion to actor-network theory* (pp. 24–33). Routledge.

2

Testing

This chapter draws together different practices of testing digital security in the companies: (desk) risk assessments, penetration testing, and incident reporting. The chapter employs topological literature from ANT and post-ANT. In conversation with the empirical descriptions, they allow to unfold different patterns of relation in which digital security is performed through testing: the regional, network, fluid, or fragile pattern of relations. The chapter ends with two implications of how an interest in topologies can offer new insights into digital security and further its ontological opening for both for STS/Anthropology and digital security research.

Testing does not simply—meaning neutrally—evaluate *how* secure a feature is, but actively performs *what* being secure means in a particular setting. It is this ontological work that I am interested in here. Testing activities, the chapter argues, perform digital security in different "patterns of relations", terminology borrowed from STS (Sørensen, 2009). Not all testing practices produce rigor and assurance, but some engender variance, even ambivalence. This re-interprets the work of security officers—who do most of the testing—as being capable of not only maintaining digital security as an abstract and rigorous standard but also engaging in consensus-seeking, compromising, and forging long-term

L. Kocksch, *Fragile Computing*, https://doi.org/10.1007/978-981-99-9807-4_2

collaboration. In the terminology of topological literature, testing practices include more "fluid" forms of digital security. But this is not without tension. Fluidity is considered "incompliant" by standards and some security experts. It needs some specification of how local variance becomes possible in digital security, which is why I suggest speaking of "troubled fluidity", which incorporates its precarious nature.

Authors have suggested different typologies of tests, for example, threat modeling, assessment audits, and effectiveness measures (Donaldson et al., 2015). However, what such typologies lack is an in-depth understanding of how *testing* is done in practice and how such practices accommodate uncertainty and divergence.

Anthropologists have shown that audits are of central importance to the allocation of accountability in contemporary organizations. Marilyn Strathern argues that audits have become central at the beginning of the twenty-first century as organizations are prompted to comply with ethical and regulatory frameworks. As part of the "audit culture", assessments steer and punish corporate actions, making audits central management operations (Strathern, 2000).

Audits have a double configuration: They involve a series of material devices and practices, and they mobilize moral order and ethics. Audits produce organizational accountability in the forms of quantitative reporting and by linking to higher goals of organizations or governments (Strathern, 2000, p. 2). Being accountable by means of audits means being able to account for, report, and make visible, as much as being responsible and morally committed.

This chapter follows different digital security testing practices in the companies, from the acts of testing to communicating results and taking organizational or technological consequences in order to comprehend digital security testing's performative effects. It does so to determine how testing locates, accounts for, and intervenes in, digital security. Specifically, I am interested in where digital security audits perform blind spots, which practices fall short of being accounted for or where audits are (partially) inconclusive and encourage adaptation and handling ambiguity. This draws together attempts in digital security research to understand testing practices (Sabillon et al., 2017) and mobilizes anthropological interests in the performativity of audits and assessments.

By exploring the empirical examples, I develop an understanding of the practical implications of conducting digital security tests, which include negotiation, compromise, and prolonged collaboration. Rather than assuming digital security assurance as a one-off checkbox, testing practices in the companies show rather complex negotiation processes, which are not well accounted for in existing audit typologies (like Donaldson et al., 2015). Particularly, such typologies often see testing as a comparison between standards and "actual" practices, whereas during testing in the companies, local actions and standards were in a more complex relation. Rather than standards on the one side and practices on the other, standards are translated partially, suspended for a while, swapped out, and slip out of focus.

Some digital security testing is messier; it does not produce stable evaluations and indicators but forges opportunities for collaborative consensus-seeking and compromise. The argument is not that such messiness should be eradicated, nor that tests should take efforts to comprise "all" facets of digital security, but rather to understand how digital security tests handle their obligatory partiality and generate stability despite variance and ambiguity.

In short, the chapter asks: How does testing perform digital security in the companies and how does it account for adaptations and divergence?

Annemarie Mol and John Law suggest that social theory in the early 1990s had engaged in comprehending two topologies of relations: First, relations were parsed into regions where entities were grouped together into homogeneous fields that could be neatly separated. Regions co-exist, can interface, but stay unaffected by each other. Digital security, exemplarily, may be parsed into technological entities and human entities, encoding disciplinary order as a consequence (Barry & Born, 2013). Secondly, ANT has introduced the topology of the network: In the network form of relations, entities are not grouped together by seemingly inherent criteria (being technology or human), but through ongoing acts of co-shaping. While it was ANT's intention to place agency in the co-shaping of entities rather than pre-assigned inherent qualities (humans can act, objects do not), the network pattern soon became problematic. Only those entities that enrolled—were turned immutable—were assigned agency. If actors (human or non-human alike) failed to comply, the entire network became at risk of failing. In Latour (1996, 1999; Latour & Venn, 2002) and Law's (1987, 2002) initial studies, this was for good reason as they attempted to re-construct how a variety of actors contributed to the success or failure of a technological innovation. From the perspective of post-ANT scholars, however, their focus on success or failure created a "managerial bias" in which strength and dominance prevailed (Law, 1999). While the network pattern of relations overcomes the purification of regions, it showed a tendency to overemphasize function and compliance. To develop ANT further, Mol and Law (1994) introduced a third spatiality of the social that they term "fluid" or "flow" (and later "fire").

The post-ANT proposal is that relations within networks become productive not only when they are rigid and aimed toward a common cause but also when they are fluid and interchangeable. Even the most rigid-seeming objects need adaptation to local circumstances. "Shapeshifting" becomes considered necessary to allow objects to travel. Technologies are a good example as they often work, not because of their rigidness, but because they can be applied to local contexts. Marianne de Laet and Annemarie Mol (2000) introduce the case of a water pump that is being introduced into different villages in Zimbabwe. The pumps could be materially rebuilt, and pieces could be dropped or added to fit local requirements. Furthermore, the flexibility of the technology allowed the pumps to create varying social relations in the villages. De Laet and Mol infer that the "bush pump" was successful not because it was the same everywhere and local villages were organized to comply with the pump, but because they allowed for variance in both the material components of the pump and the social relations it forges (De Laet & Mol, 2000). Consequently, they contend that technological objects or socio-technical associations are not only stable when they are kept together and everything stays in place, but also when objects take a fluid shape and are stabilized by plasticity and gradual modifications.

It misses the point to speak of the bush pump as a "failed network" because material compositions of the pumps differ, and social relations are built differently through "it". Law and Mol argue it is more useful to think of the bush pump as a "fluid object" that retains its shape as it flows but does not disintegrate into completely different bush pumps either

(Law & Mol, 2001, p. 6). It is important that fluid objects change shape, but they do so gradually and incrementally; from time-to-time bits fall off and are patched on. The pumps are different, but they still allow for continuity. There is a familiarity between the pumps, one that is not a defining feature—a stable network—but enables partial overlapping and resemblance (De Laet & Mol, 2000; Law & Mol, 2001). What remains across all pumps is the function to pump water and to forge meaningful relations in the communities.

Fluid patterns do not break down nor are they dismembered by these deviations; fluidity is what keeps them together. A rigid pump that could only pump water in one way and can only produce one type of relation would have failed. The pump remains the same while its component change, that is, it remains a device to pump water but is comprised out of different material and social relations.

Mol and Law provide a similar example with the diagnosis of anemia through diverging practices:

> [...] in addition to the laboratory, there is a clinical way of screening patients and finding anemia. It makes use of experience rather than accuracy. So, the clinical gaze doesn't create sharp numbers. But it certainly produces diagnoses. Effective diagnoses. Diagnoses that work so well that the absence of a hemoglobin test may even go unnoticed. (Mol & Law, 1994, p. 654)

With that example, the aim is to point to the intersection of specific target values for identifying anaemia in patients and the practical, often creative ways that doctors apply. A disease called anemia is still diagnosed but in different ways. Law and Mol write:

> In a fluid space normativity is gradient rather than a cut-off point. When they deal with anaemia, tropical doctors aren't trying to reach a specific threshold, but to nudge the Hb level in the right direction. To push it up a bit. [...] In a fluid space it's not possible to determine identities nice and neatly, once and for all. (p. 660)

In other words, resilience of what anaemia is, is gained through difference and gradual adaptation rather than through conformity and invariance.

But this comes with a warning that shifts that are too abrupt and changes that are too fast can lead to failure. Fluid patterns are fragile. Harmonization or standardization is just as much a potential threat to fluid spatial forms as overly abrupt and swift changes (Sørensen, 2009). In her example of a virtual learning system, Estrid Sørensen argues that pupils employed various practices to make use of the system, often in contrast to what educational experts and software designers had intended. She points out that what scientists, teachers, and educational experts see as a failure of the system, may not be seen that way by students, referring somewhat inexplicitly to the fragility and tensions that fluid systems can entail. She notes that each pattern of relating she analyzes is also troubled in a way, for example, lacking representation or collaboration. As Mol and Law stress: "[…] fluid spaces are no better' than regions or networks. They ar eno more attractive. Or virtuous" (p. 663).

The patterns of relation provide a visual language to inquire into the spaces (topologies) constituted by security testing practices in the companies. The point is not to identify in which pattern digital security is "best" performed, but to unfurl the various topologies performed through testing. Through applying the trick of topological thinking, including the juxtaposition of patterns and thereby widening conceptual repertoires, allows to inquire into the ontological work done by security testing.

Digital Security as Regions

Company Branches

At the time of the study an ongoing discussion at Powino was *where* to invest more in security. A security officer conducted a risk assessment to measure the potential monetary loss of a cyberattack in different sectors of the company. Powino consisted of three main branches that had several more sub-departments. The main sectors were grid and infrastructure (including renewable energy sources), digitalization, and customer relations. His analysis showed that potential loss differed between the sectors:

Customer relations could in a worse case result in a loss of about 100 million euros, while grid and infrastructure could lose up to 1 billion euros. Defining monetary loss was a strategy to decide *where* security incidents would cause the most harm, measured in monetary values. In that way, security was mapped in different sectors or regions.

Where to invest, however, was not only a decision based on what sector was most vulnerable but also what security measures promised the greatest effect. Each region carried a different security objective. Security in grid focused on strict access controls and cyber-physical control systems (e.g., access to buildings). While in sales, customer data was protected by strict classifications of what data was more "critical" when lost, that is, how much reputational damage it would cause or how much in penalties could be faced according to the General Data Protection Regulation (GDPR). By testing for the monetary damage and hoping to thus locate where security efforts were most required and most promising, security was mapped out. In grid, it meant ensuring access control, while in sales, it meant detailed data classification. What to make of this difference? First, it is important that this test mapped out security by suggesting that branches are homogeneous entities. On the map of these three regions, security in the sales department is less important than in the grid department.

Testing where the monetary loss would be biggest combines digital security with other values, for example, high or low cost and "good" or "bad" reputation for the company. How to locate digital security was a question of in what region security was most needed, and *what* security was needed. It was a question of what security would be needed in the future. The underlying question was where would a security budget prevent the most monetary harm? This question was related to for whom it would be best spent. The security officer makes a clear point that Powino is only responsible for the failure of the grid but not for consequent damages, for example, to hospitals or the stock market when the power is off. He makes it clear that "Other companies are responsible for that". Thus, there are strict boundaries between regions of the company and other companies/infrastructures, that Powino is not responsible for.

The risk assessment makes branches and their security comparable by calculating how much money could be lost. Thus, we can define the first region of security as different branches of the company.

Countries

Another form of the test was conducted by Powino's security department in sending out emails that resembled "phishing e-mails". Phishing emails contain links or attachments with malicious code that, when clicked on, are automatically downloaded to the computer. The security department of Powino tested response rates to those phishing emails by mimicking the design of common phishing attacks. Instead of hiding malicious code, however, the links would lead to websites of the IT department containing more information on what phishing attacks are and their potential harm. The officer who explained that to me elaborated that the decisive indicator to measure how aware people are of phishing attacks is how many people click on the link. The *"numbers have improved from 20% to 5% in some countries"*, he explains, *"while they are still quite high in others"*. Powino had different offices in several European countries besides Germany: the UK, the Netherlands, Poland, and the Czech Republic.

The categorization of countries into "low awareness" and "high awareness" performs a regional logic. Awareness is defined as homogeneous within the region by setting a fixed "cut-off point" (Mol, 2003, p. 60) of 5%. According to regional tests, security in the sense of avoiding phishing emails is reached when less than 5% click on a phishing email in a test. Such targets serve as indicators for security, and as Strathern (2000) has noted, those are notoriously partial and have political consequences. However, in this example, that ambiguity is not dealt with.

The officer sets a "target value" for security of 5%, above which actions become necessary. This test locates security according to geographic regions instead of per branch in the company. Here, security is *less* in countries with higher response rates. It is *more* in countries with lower response rates. If security *is* lower response rates, it can be found in one country more than in another. If security is a monetary loss, it can be found more in one branch than in another; if it is response rates, it can be found more in one country than in another.

There is an additional overlapping of regions here: While the tests compare between countries, they also aim at the heads of employees as regions. When indicators are as high as 20%, awareness campaigns target individuals and thus attempt to lower response rates in countries. Thus, countries and heads come together in this test, one being the specification which heads require targeting. But the test does more than measuring; by linking to the IT department's website that gives further information on phishing it actively engages with employees and performs an educational task: If you have clicked on the email, you should learn more about phishing.

Such an approach again has problems in practice. Saying that security is bigger in one country than in another prompts the question: Where are the boundaries drawn? Saying that one country is more secure than the other is certainly not true, as the tests are conducted within the company and not with the general population. In practice, there are complications in the regional pattern. What about the domains these national branches work in? When those with higher response rates work in Sales, it is perhaps less of a problem. Interestingly, phishing experiments were highly controversial in the companies as mimicking phishing attacks was considered unethical because it employed tactics used by attackers, and some security officers felt that deceiving employees could potentially harm the trust between security departments and employees.

In addition, it is noteworthy that conducting digital security assessments in that form is done more in some geographical areas than in others (Sabillon et al., 2017, p. 254). Such regional mapping of digital security along with a clear distinction between "human" and "technological" regions in digital security marks the dominance of regional approaches in digital security.

Heads

Consider where Simon, one of a handful of security officers at Powino, sees digital security located: To make technologies more secure, he says digital security must be *"in people's heads [...]"*, and he adds: *"It is not in people's heads yet to do security by design"*. "Security by design" refers to considering security very early on in the process of designing a new technology (Kreitz, 2020). What Simon aims at here is the "awareness" of employees at his company. Various testing practices are employed to assess the awareness of users for digital security matters (Rahim et al., 2015).

To this security officer, the way of making security better is by getting "into" the heads of people. This seems like an odd realization. As digital security is a technological problem why would it be found in people's heads? What he means is that people must remember to consider security when building new technologies. They must make room for it in their planning of development steps for technological products. They must take the time to test, pilot, and consider security when establishing functions. To the security officer, this starts by getting security in the heads of people. In the heads, however, security was notoriously hard to test. Note that the phishing experiment above has tailored heads in a way by testing the "awareness" of phishing attacks among branches. Whereas the argument was that through the phishing experiment, particularly homogeneous areas of identified concerns around awareness foreground that digital security is also seen as a matter of individuals' perceptions or "mental models" (Bravo-Lillo et al., 2011).

Security officers at Powino and Covy often complained that security standards were one thing and getting security "into the heads of people" another. They drew a similar line when arguing that technological security was easy to reach, but making humans comply was much more problematic. The relation between humans and technologies is seen as a problematic one since humans endanger technological security. While material technologies could be adapted and were perceived as "soft" as they could be adapted to security requirements, humans were considered more "stubborn", incontrollable, and unruly.

It is not strange to me that he said security must be introduced to the "heads" of people here. "Awareness campaigns" send out flyers or convene workshops that remind people to watch out for security-relevant issues. They teach them to be aware of emails that contain links that can introduce malicious software to their computers ("phishing"). I have received paper notebooks and pens with security advice: "*Tidy up your desk to avoid unwanted views*", "*Always use [Windows]+[L] to lock your computer when you leave it unsupervised*". The head is a popular target for digital security campaigns and tests. The awareness campaigns make use of daily encounters to make their warnings more realistic. They suggested that best practices and routines prevent digital security issues. Awareness is a major objective of digital security research as well, one that often explores the awareness of specific homogeneous fields, for example, geographical regions, individual companies, or social groups (Bada et al., 2019; Tirumala et al., 2016).

Targeting awareness and thus "heads" follows a humanistic view of experience and creates the trope of a "deficient user", as has been criticized (Klimburg-Witjes & Wentland, 2021). It argues that digital security can be targeted by urging users to make better choices. This follows a "logic of choice" where individuals are considered isolated entities, and actions rational consequences from mental processes, made on the available information (Mol, 2008, p. 1). To an STS reader, this jumps out as a humanistic understanding of agency while it emerges as a target for assessment in digital security.

The third region that can be identified are "heads" that are targeted by "awareness" campaigns and reminded by material devices to make secure choices. Following are definitions of other regions where other forms of digital security can be found.

Technologies

What other regions were identified by digital security testing? Assuming that devices and networks would be a good starting point to locate security, I asked interlocutors for an overview of systems and measures to which they usually responded: *"Yes, that would be good to have"*, joking that this could be my task to map out all systems and data in use. The network of systems in companies like Powino and Covy is complex and indiscernible to a remote observer, it is hard to map out "from above". We could call this a "problem of scale" (Mol & Law, 1994, p. 642): Where is security to be found in the unknown network?

It is not easy to locate security from "above" and difficult to pinpoint it "within" machines. Interlocutors sometimes called it out to me when they were busy on their laptops and a password was required: *"this is security"*, they would say, often suggesting it hindered their current workflow.

Within those volatile moments, security could be defined: a password that was needed, a cupboard that was locked, an encryption checkmark that had to be set.

During penetration tests, a series of techniques is employed to actively breach a system. Penetration tests allow to see a system from the outside (like an attacker). Penetration testing is also known as "ethical hacking", as hacking techniques are used to report possible entry venues that can then be fixed by developers. The intruder ("tester") locates security "within" the system—performing it as a homogeneous inside that stands in opposition to a clearly defined outside. Testers switch roles from defending to attacking purpose to see the system from the outside. By doing so, they perform the technology as an isolated region in which attacks can occur based on insufficient code or unintended network combinations.

For the moment, it is important that it is possible to create a map of digital security where different regions coexist and are enabled by testing that separates different regions by defining homogeneity. We could call it a "homogeneous field" as there are strict boundaries between regions where security is abundant (in technologies, in some branches, in some countries) and those where it is scarce (in heads, in other branches, and other countries). Following Mol and Law, digital security is thus mapped in a space that is "exclusive". There is "neat division, no overlaps" (1994, p. 647) between the regions. Regions can stay remote; they can be targeted with particular measures.

The previous examples demonstrate tests in which digital security is identified in certain regions. From this follows a comparatist logic—one region can be compared with another: humans are less secure than technologies; country A is more secure than country B; branch 1 is more secure than branch 2. In the matter of regions, digital security becomes comparatist.

As Sørensen (2009) notes, regions are not only separated by geographical locations but also saturated with norms and power. In her example of a classroom situation, the region of the teacher (her desk, blackboard, chalk, etc.) and the region of students (their desks, pens, etc.) are distinct, even if one can be invited into the other. Asserting power over company branches separates them from the headquarters in the center, perceives of heads as less compliant, and performs normative guidelines of how to act and do security.

The regional approach enacted in some tests entails distribution of the object to scattered locations: Digital security is isolated to heads, technologies, company branches, and countries. In addition to Mol and Law (1994), I want to emphasize that regional digital security operates along the lines of a specific allocation of responsibility for a potential error or breakdown. When digital security is approached as isolated in regions, regions become responsible for possible errors or failure. The responsibility for security in a regional spatiality lies within entities: specific areas of formal accountability. Such entities can be the ones above, i.e., organizational units or regions that are isolated from one another (heads and technologies), but can also be perceived more generally, that is, “users”, “developers”, “companies”, and “law”. As Law (2000) illustrates in his analysis of the Ladbroke Grove train disaster, the regional space assumes a flat surface that is divided into isolated territories neatly separated. Asking the “where question” in such a regional logic means pointing to a distinct territory or arguing about the right territory (p. 13).

The notion of approaching digital security in isolated regions is obviously simplified—the office themselves attested to that as I will illustrate in a minute—but it picks up on something important. Responsibility tends to be allocated in regions in digital security: technological error, human error, weakest links, etc.; all common notions that point to the simplification of responsibility in digital security. I want to develop another argument here by looking at different spatial forms. It is often said that *“this organization”*, *“this technology”* or *“this user”* caused an insecurity. Such language is about locating responsibility in a definitive region. And there are various techniques that have been developed that address isolated regions in digital security: psychological measures like training and awareness campaigns tailor people’s heads; technological measures like improved encryption or access control tailor toward technologies; organizational measures like risk management or phishing campaigns aim at company branches and countries. In the pattern of regions, the entities stay isolated, and can be identified and targeted as responsible for security. Regions are not capable of noting any far-reaching relations between entities, as they approach each of them in isolation, sometimes by putting them up for comparison—which ends in descriptions of either region (heads, technologies, organizations, countries) as inferior or superior. Measures in the regional pattern are individually cascaded as each entity is considered in isolation. This is also represented in literature on assessments that acknowledge the difficulties of testing digital security because it spans across technological and human actors, or across regions (Al-Matari et al., 2018).

There exists some dispute about what region is more important to tackle in digital security research and which one should not be made responsible without any consideration. This stays somewhat in a regional logic. In such disputes, it is often brought forward that users and developers should not be blamed without understanding their concrete contexts, power asymmetries, and working procedures. This adds another spatiality. Identifying security solely in regions runs into problems. Some branches have higher security requirements than others and are not easily compared to each other. Technologies are reconfigured in practice, reshaped, and adapted. Running phishing experiments can be considered a violation of trust relationships in companies, etc.

This suggests, it needs an understanding of the relations, "contexts", and connections between branches, technologies, humans, and so forth in order to define digital security problems. In the next section, I will attend to tests that perceive digital security not as isolated regions but as a *network* of different controls and actions that must align and be disciplined in order to achieve digital security. Applying the trick of ANT here, simply suggesting a different spatial pattern can come in handy to explain such inconsistency. It allows us to consider how tests identify security beyond entities as a matter of relations.

I have asked above how security testing performs different patterns of relation. The first testing practices created digital security in a regional way; digital security was parsed into homogeneous regions that could be arranged and compared. Testing enacts this regional form by finding security "more" or "better" in one place than in another, or addressing specific regions because they show "deficits" in comparison. This mode of testing not only produces regions but also produces navigation of where to intervene. Regional security can be tackled by jumping from region to region, attending to one issue after the other.

Digital Security as Network

In the network pattern of relations, testing performs potential incidents as the product of a failed network of controls where individual components or actors lack harmonization or compliance to the rest of the network. In this pattern, failure results from a lack of successful translation into concrete actions. Digital security is performed as a tightly knitted network of human, technological, and organizational measures. This is inclined to look for responsibility at the hands of managers or those doing overall coordination work.

The network pattern suggests a mechanistic idea of digital security where every bit must fit together. Stability is reached through making digital security mechanisms and practices the same everywhere in the company. The object of stability is then not an isolated region (that should at best be kept out of complex interactions) but the creation of sustainable relations: how well does the network of security mechanisms work together?

We can employ Mol and Law's definition here in a direct way:

> In a network space, then, proximity isn't metric. And "here" and "there" are not objects or attributes that lie inside or outside a set of boundaries. Proximity has, instead, to do with the identity of the semiotic pattern. It is a question of the network elements and the way they hang together. Places with a similar set of elements and similar relations between them are close to one another, and those with different elements or relations are far apart. (Mol & Law, 1994, p. 649)

Network refers to a conceptual topology that is opposed to the Euclidian space of coordinates (places, smaller, bigger, etc.). Network also refers to a space of connection, disregarding geographical place and geometrical size (Latour et al., 2014). Networks—in the ANT sense—consist of only one level, while Euclidian topologies—that are present in the spatiality of regions—consist of at least two layers: micro-macro, parts-whole, branch-corporation, heads-organizations, country-company, and so forth.

What the network pattern shifts toward now is understanding not how individual variables contribute to security, but how various variables play together. It shifts attention to relations between indicators. Instead of focusing "where" security should be targeted—thereby limiting its possible intervention to entities in question—testing activities aim at identifying how different entities act together and thus *become* in/secure. This shifts the logic from individual entities to a matter of relations, whereas entities become in/secure through adding items to their connections.

In the network pattern of relations, learning what an entity is (a security control, a user, etc.) is done by adding other entities to the list. The precarity of one entity becomes clear through relating it to other entities. Using a new piece of technology can become a problem, when it is related to "old" software that includes other security policies, potentially overwriting each other. In a way, approaching digital security as a network pattern demonstrates that the "whole world can be grasped" (Latour et al., 2014) through the security point of view where each entity then becomes relevant in relation to digital security.

ISO Standard as a Network of Security

Digital security is standardized by the 27 K-row standards[1] for information security that were developed in collaboration between the International Organization for Standardization (ISO) and the International Electrotechnical Commission (IEC). The row consists of 50 standards covering areas as preparedness measures (how to avoid digital security incidents); incident management (how incidents should be reported and processed); and forensics (how incidents should be analyzed and prevented in the future). The set of standards was originally

[1] Extensive work has been done in STS on the situated histories of standards (Lampland, 2009), as well as on their local and specific performance (Hanseth et al., 1996; Berg & Timmermans, 2000; Timmermans & Berg, 1997). This could have been a valid way for this book to trace the specific sorting and categorizing done by the ISO standard with relation to literature in STS (Bowker & Star, 2000). But in order to stay somewhat open to different enactments of digital security, standards are only one form in which digital security is done—one that is persistent—yet not exclusive. As this chapter indicates, stability is reached in different ways in digital security, translating the standard being one among others.

published in 2005 and updated in 2013. In 2019, it was reviewed and confirmed.

When referencing the standardization of information security, interlocutors most prominently mobilized the first standard in the row: 27001, with the full title: *ISO/IEC 27001:2013—Information technology—Security techniques—Information security management systems—Requirements* together with its annexes that specify 114 security controls. This first standard particularly encapsulates measures to prevent malicious access to systems and information. The 114 controls entail technical measures, such as password encryption and organizational measures as to how to organize segregation of duties and how to conduct regular tests to align security controls.

While an official certification can be obtained from certification authorities (e.g., the German TÜV), and both Covy and Powino had certified infrastructures, the 114 controls also work as a guideline for those infrastructures that are not certified. Even those systems that were not officially certified to comply with the standards used the 27001 Standards as a guideline to develop sound security concepts. Through tests and assessments, the ISO standard has far-reaching relevance even when no certificate is sought for.

The standard family is intended to be universally applicable: "The requirements set out in ISO/IEC 27001:2013 are generic and are intended to apply to all organizations, regardless of type, size or nature".[2]

The ISO standard is the most prominent digital security standard, even though others exist. Critical infrastructures like Covy and Powino are legally obliged to gain an ISO certificate and introduce certain measures into their products according to the German "IT security law".[3] During interviews with security officers, they often reflected on this standard in particular and it acted as a constant reference that security officers brought up in conversations.

The most important aspect of the ISO standard was the risk management system, or Information Security Management System (ISMS). Risk

---

[2] https://docs.microsoft.com/en-us/microsoft-365/compliance/offering-iso-27001?view=o365-worldwide

[3] https://www.bsi.bund.de/DE/Das-BSI/Auftrag/Gesetze-und-Verordnungen/IT-SiG/2-0/it_sig-2-0_node.html

management covers a wide range of possible risks that involve physical access to information. The ISMS specifically approaches digital security as being distributed across human and nonhuman actors as it suggests listing risks that are introduced by both technical mechanisms and organizational, architectural, or social factors. While the former is reached by testing software code or implementations, for example, penetration tests, the latter requires organizational interventions, for example, by implementing segregation of duties or code reviews, aiming at enrolling actors in the goals of the ISO.

The description of the use of the ISMS points us to a shift from isolated human or nonhuman regions where digital security is relocated in a more comprehensive network of security measures that must be taken. The ISMS approaches security as a network of technological controls, awareness campaigns, tests, audits, and risk assessments. Rather than pointing to isolated regions, the standard encompasses a comprehensive framework that requires 114 security controls to be in place in order to achieve a stable digital security network.

ANT's concept of a network agrees that through various "trials of strength" (Callon, 1984), various actors are enrolled in a network that in turn through the enrollment of actors gains stability and continuity. Neither precedes the other. Meaning that without the actors, the network would fall apart, whereas without the network, actors do not have their specific relevance and become actors in that sense. ANT scholar Michel Callon explains why the notion of enrollment is used:

> Why speak of enrollment? In using this term, we are not resorting to a functionalist or culturalist sociology which defines society as an entity made up of roles and holders of roles. [...] [Enrollment] designates the device by which a set of interrelated roles is defined and attributed to actors who accept them. (Callon, 1984, p. 74)

In ANT, enrollment does not entail only human actors, but rather technologies, devices, political agendas, etc., or a whole assemblage of entities that are needed to make a network stick together. It, therefore, shifts attention to how actors are enrolled and become associated which allows a network to be spoken of in the first place. Each relation is tested to ascertain whether it holds in place to become reliable and gain relevance.

Translating the Network

In September of 2017, I participated in the assessment of a new platform (the "*Data Lake*") at Powino that was at the time being developed by two architects. The *Data Lake* provides a data storage and analysis platform for large-scale data by partially employing cloud technologies. *Data Lake* infrastructures were introduced at Powino and Covy at that time to allow the use and analysis of large data masses that operate on easy-to-scale servers. The lakes were also used to facilitate more centralized storage of company data and thus new ways of combining different data sets (on the epistemic consequences of this introduction, see Chap. 5. The *Data Lakes* were part of a bigger transformation in the companies toward basing corporate decisions on the results of data science.

The assessment was conducted by Simon, one of Powino's information security officers. Several revisions were required along with at least two two-hour phone calls in which its results were individually discussed by Simon and two *Data Lake* architects (Georg and Anton) from the data analytics team.[4] Based on the 114 controls provided by the ISO standard, Simon had developed a set of 355 specific items that were relevant to IT systems at Powino. He then selected 23 of them as relevant to the *Data Lake* project. He used these more concretized items to assess the documentation of the *Data Lake* according to whether they complied sufficiently with what he interpreted from the ISO and translated for the concrete case.

By translating the 114 controls, Simon added specificity and already considered what controls are relevant for Powino. His fear was that while one system alone could be secure, as soon as it was integrated with another system, problems could arise.

[4] While at Covy, the Data Lake was built by data scientists (along with what they called "the data office"), at Powino, the team was called data analytics. So far, I have found no empirical distinction between the two teams. Thus, I will use the terms "data science" and "data analytics" in the way they were used in the field acknowledging that in my observation, they meant the same thing.

This was particularly pressing as some data that was currently held separated in storage units could potentially be brought to the new infrastructure. Data that was "safe" in its current system could become less safe in the new one. What changes was not the data but the particular relations that the system created, for example, how it generated access to specific parts of the system?

The report specifies that access is automatically given to data that lies within the criticality levels between one and three. Criticality levels determine how confidential data is: one means data is public knowledge (for example, reports of market developments); two means data can be shared within the company (for example, knowledge about a new product); three means data is circulated only among particular groups of people within the company (e.g., customer data); four means data is highly confidential and cannot be shared within the company. Only for level four data is access control required to "explicitly allow"—whereas for other types of data, "explicitly deny" is sufficient. The latter encompasses that all data is accessible by default and users must be explicitly denied access. The former, however, does not allow access by default and access must be explicitly allowed. The ISO standard argues that explicit allow is preferable because no data become "accidentally" accessible. What Simon's assessment suggests is that data or users are not essentially more or less secure (measured with an external threshold), but that through specific relations established by the access control system data that is critical, data can become accessible, which would be a security problem.

Simon suggests that another aspect of such relational securing is not to address specific users to enhance the security of the system but to assign access rights to particular accounts—which do not necessarily represent individual users but could also be machines or one user having several accounts. Rather than giving one user several access rights, accounts could be created for specific tasks, without general login rights. This speaks to the idea of digital security as stretching across various entities, not limited to human, technological device or otherwise. Instead, through the enrollment in the security network, entities become the capacity to act as entities. Furthermore, only very few accounts are allowed to delete data, which allows data to be accumulated.

The report also states that while accounts could be configured correctly, they might become a security problem over time, and a regular review of access rights is necessary, as much as a process to revoke rights when projects end, or when new data is uploaded to access repositories. This again emphasizes that neither data nor accounts are essentially insecure but can *become* a problem when their relations change. Accounts must also be able to expire over time when the relation to real people changes, e.g., when people leave the company or retire. If an employee that leaves the company could still access data, particularly data that is considered critical, this becomes a security issue.

Finally, the report also emphasizes that such constellations become problematic in which data is not officially listed, for example, when it lacks indexing in the list of company assets. Not knowing that data exists hinders a proper overview, and thus hinders monitoring and control.

Security in the network pattern is contested when one failure exposes the rest of the system. Because each insecurity is considered a "weak link" in the network, tiny mutations can endanger the integrity of the whole system. Then, not just the new technology is in question but bigger networks like the company infrastructure or Germany's (potentially even Europe's) energy supply.

In this way, security is not a matter of isolated regions, but of connected networks and dependencies between smaller and bigger systems. In these examples, security is not definitive in one territory of a flat surface as it was before, it is a matter of a network of human and non-human actors that depend on each other, and anyone can risk the failure of the entire network (and adjacent networks).

Simon's assessment proposes a trial of different relations. He went through the documentation of the system, testing each relation for how strongly it might hold in the face of an attack or how it enabled unwanted functions, e.g., access. In the terminology of ANT, this can be perceived of as a "trial of strength", a contest between two opposing forces with a robust network emerging in the end (Latour, 1990; Callon, 1984).

The trial proceeds in different phases. The first phase is to translate general standards into locally relevant controls. The second is to compare the controls to existing applications in the system and conclude how much they differ. The controls thus emphasize the entire network by looking closely at specific items. The third phase is to articulate specific changes in local controls that allow the stability of the whole network and also refer to the standard. This double relevance is important as the stability of the system is inherently linked to the considerations of the standard.

The ISO standard and Simon's efforts to translate it into concrete controls for Powino transpires digital security as a tight network of controls that must all be in place to ensure stability and continuity in digital security. The standard is generalizable as indicated by its exact wording above; it "applies" to all kinds of organizations, permitting reliable and stable security mechanisms.

This illustrates an important shift from the previous notion where digital security could be targeted in isolated regions to be reached. The network pattern emphasizes the relations between regions and defines their in/security not by looking at them in isolation or comparison, but by relating them to each other.

Testing for the security of relations, Simon also notes that it needs a good understanding of people's relations to each other. Building digital security as a network requires building strict relations between people, e.g., by controlling each other's actions. He explains:

> *For example, applications that transfer money. If one person can make changes, she can transfer one cent for every transfer the program makes to her account. Nobody notices, you can even correct it later. So, if only one person oversees this, this is a great thing. We want to prevent that. But if you say, ok, it's two people from the same team—is this more secure? Before we had the rule that a developer and an administrator had to work together, they are in different teams, and this is enough to comply with separation of duties—Does it suffice if they sit next to each other from the same team? This is the interesting question if you can speak of a separation of duties if they have the same goal, that is to make the software work. You don't have control anymore.*

He describes here a shift of composing teams in a way that administrators are now often part of a team and might share the same goals as developers, e.g., to finish a product for a specific deadline. It is more difficult now to assume that since they are in a relationship that one can control the other. Simon needs to pay attention to the relation between developers to make sure security (here separation of duties) is possible.

In addition to the testing objectives introduced in the previous section: psychological awareness, technologies, and organizational sectors, this section has introduced additional ways of attending to digital security, i.e., testing relations. The ISO standard and the assessment translated from it are not tailored toward individual components, as were the measures in the previous section, but are attuned to a functioning network. Assessments are capable of understanding concrete relations between entities that unfold in a system such as the *Data Lake*. In the network pattern, security measures aim toward the relations between entities, not the entities alone. This resonates with an ANT understanding of networks that what is important are the shifting relations between entities, from which they emerge as relevant actors in the first place. For the *Data Lake* assessment, this meant that Simon had to understand the specific actor–network the Data Lake suggests and align security measures to it. Instead of taking a perspective from above: looking at regions from a bird's eye view, the network pattern gets involved with concrete requirements of systems such as the Data Lake. It traces ways in which attackers would maneuver through the relations of humans and nonhumans to breach into the system. The network pattern demonstrates how formerly separate regions come together and their relations are put under scrutiny. In the network pattern of digital security, failures can be perceived of as a problem in a distributed system in which the smallest "weak link" can end up compromising the security of the whole (Stachel & DeLaHaye, 2015).

Testing digital security as a network, entities—developers, new technologies, etc.—do not have attributes by themselves but gain them through specific relations. The relations are what needs attention; their explicit power dynamics, constellation with others, etc. Mobilizing spatial patterns then asking where insecurity emerges is not in definitive territories but in a web of relations. Secondly, digital security as a network insists on the need of a coordinated design, something holding the oversight and the network together, e.g., the assessment. When the object of digital security is stable relations, the question is how well they hold up and stay the same. I do not want to argue that assessments and strong relations in digital security are not needed, but I want to open the realm of theorizing a bit more by inviting conflict, ambiguity, and uncertainty into the network.

So far, security experts might still agree with this analysis that security needs not only to look at specific entities but also at how those *become* insecure when put into relation with others. They would also agree that because relations are changing and vague, it needs strict testing and assessments. The network pattern foregrounds the importance of what is doing the translation work and holding things together. Specifying the "where questions" in the network pattern can only be answered by in between and in context; and by pointing to what stabilizes relations and contexts.

But Stefan, another security officer explains that this is something of a misconception:

> *There is no central clearance anymore, and no shifting risk to the central security office. This is a shift, because currently, everyone thinks there will be a central clearance by security, that we would always do a final checkmark, and also that if they tell us they are free of accountability. That is not the case. The idea is that they become accountable and themselves say, ok, this is security-critical in relation to the standard and to the company system. We can provide methods and tools, and they can always contact us, but to scale up they need to do it themselves. This is what I learned over the years. It does not help to do code reviews myself; we need to engender people to do it themselves.*

I present this quote here because it highlights how the network pattern shifted from individual decisions (that can be "good" or "bad"), toward engendering the capacity of identifying security issues among the network; encapsulates a different distribution of responsibility. Although no entity is responsible on its own in a network, responsibility is carried collectively for the functioning of the composite. Stefan's quote also offers a hint of morality and resignation because he does not see that the ideal has been (already) reached, something that renders the entire network less stable.

So far, the chapter has answered how digital security is tested in two different ways in the companies: by regional tests producing isolated actors and factors to consider and improve on; and by testing relations and asking how they must be enrolled to increase the stability of the network.

Digital security experts can hopefully still agree with the analysis so far. But they, as Stefan did above, would also point out that the story is often less harmonic and straightforward as I have made it out to be so far. Regions are not always clear-cut, people and technologies do not always comply in their actions, and relations may be too unstable to account for a solid digital security network.

As the last quote has indicated very clearly, some tension exists between how digital security officers imagine digital security to be and how it is carried out in practice. The network spatiality cannot fully account for these tensions and pushing and pulling the edges. Hence, it requires a different description to account for the edges and ends, where the network is not as firm as Stefan would like them to be.

In practice, the network fails more often than a security officer would wish for. What happens is a familiar story of a "failing network", that is often disrupted, imperfect, and a compromise, or more "fluid" as Mol suggests (Mol & Law, 1994; De Laet & Mol, 2000; Law, 2000). In such a fluid space, both network and regional patterns appear but they are constantly in flux. In other words, there must be constant shifting between regions and networks. And hence, what Mol and Law suggest is that social forms are neither regions nor networks, but both in dynamic constellations. Approaching digital security as a region or a network does not account for the friction and constant dynamics between regional and network patterns, nor it does account for digital security's particular precarity.

To ease into the next topology, let me refer once again to examples in ANT. ANT scholar Madeleine Akrich (1992) suggests that technological objects make their users a premise to "scripts" that are "inscribed" into them by developers and experts. Technology use then is a form of "de-scription", of employing the script in practice. This puts the "inscribed" visions of the world of designers and developers in the center. Tracing how such de-scription can fail when developers know too little about the context of use, Akrich points to the lack of flexibility of such scripts.

Comparing that to digital security paints a familiar picture: While digital security regulations are inscribed with regulatory rigor and technological specificities, they may not always be de-scripted in the way anticipated. It thus needs careful attention to what is inscribed and what vision of deployment is mobilized as "technically devised inscriptions". By turning digital security into a matter of description (of using technologies "as intended"), it is delegated from experts to amateurs (who are professionals in their own field).

What remains unclear from such ANT approaches is whether a "strategic vagueness of scripts" is possible and desirable (Suchman, 2006, 193). Authors have employed the term "local universality" to emphasize that universality may not be a transcendent quality or universal "applicability of laws and procedures", but instead an effect of binding heterogeneous practices together into a seemingly universal form (Berg & Timmermans, 2000, 31). As Lucy Suchman (2006) writes: "Every form of stabilization includes, irremediably, the existence of instabilities" (p. 196).

Instead of trials of strength, Fabian Muniesa and Domique Linhardt (2011) observe "trials of explicitness" where, instead of testing networks for their robustness against deviations, they afford robustness through being rendered explicit in deviation. In Muniesa and Linhardt's contention, the gray areas in this concrete work become clear where new problems arise. Applying standards thus does not resemble a "simple" translation as attempted above but becoming involved in and affected by "explicitness"—the material and concrete intricacies of practice. Such openness to ambiguity in the literature points to digital security as a more recursive process, in which standards are subject to constant re-specification. In order words, the next section aims at answering the question, how is the performativity of testing challenged when entering "the wild" (Dourish et al., 2004; Hutchins, 1996).

Digital Security as Fluid

I have noted above that the network pattern that is performed in tests mobilizing the ISO standard shifts attention from isolated entities to translating standards and enrolling actors in a stable network of controls. But the network pattern assumes rather stable associations or relations: once things are enrolled, they must maintain their position, or the system will fail. As suggested by post-ANT scholars, some networks do not persist despite defiance and deviation but because of them. The following section portrays testing that produces digital security with disorder, variance, and fluidity. Incidents in this pattern are not the result of too little harmonization, but, potentially, of too little adaptation and adjustment. Fluid patterns of relation fail if they are too harmonized. Accordingly, these tests perform security in the form of "good" adjustments.

The security assessment introduced previously was accompanied by two phone calls involving the security officer, two architects and me. The procedure was to go through each item in the assessment and discuss its details and possible consequences for the "*Data Lake*". The purpose of the meeting was both to ensure that the items were brought to the attention of the architects but also gave a change to discuss what to do to achieve the security as desired by the assessment.

The process was intense and lengthy. During the first phone call, it was decided that a second would be necessary because only 9 of the 23 items were covered. No one had anticipated that going through the issues would take this long, and during the call, it was often jokingly celebrated ("We made it to issue number x!"), subtly denoting that it was not an easy process to discuss the consequences of each item. It took intense negotiation and going back and forth between items to apply the list of findings translated from ISO standards into concrete development tasks for the team.

One of the architects read out the issue description and explained what they understood and whether they agreed with it. The security officer then explained the reasons for the issue, sometimes needing a moment to remember what it was about. He would then explain that something was *"required by the security guideline"* or that *"people in the data center do it differently*", comparing the existing documentation with what was the standard way according to the ISO or best practice at Powino.

Two issues were solved by one of the architects even before the assessment had taken place, and two further items had become obsolete because the architect had dropped specific applications in anticipation that they would not comply with the security standard. In other cases, it was not the technology that was adopted but the documentation on which the assessment was founded had not been updated. In the latter case, the security officer would say: "*Then write it like this*", insisting that the documentation more appropriately reflected what was in the development.

If the documentation was not unspecific, outdated or the issue obsolete, the test results were discussed. This happened several times.

> *This is probably a bit of homework for you. We* ***do*** *store it [account information and passwords] in different ways. We have different authentication mechanisms and we have implemented a key rotation; demanding password changes every 90 days. Probably not everything will be done automatically, but we will try. But you also want the key to rotate?*

The architect here explains that they have already implemented a rotation in password setting, and separate and thus secure storage for account information and passwords, and that he does not understand why even more is wanted from him.

> *The security officer responds to this by widening the issue: "[short break] I remember now. Service accounts. Where are the passwords stored? Who in the team has knowledge of the passwords? If you say, ok, we store it there and access from team to passwords is locked. Please write this down"*

In this case, the atmosphere in the phone call was tense. The architects defended their system and argued that the assessment's results were too strict or "on top of" other measures already built in. This was solved (temporarily) by adjusting the document.

However, it was also noticeable that the architects actively reflected on their work by adding that they might not be able to do everything automatically but would "try" their best. One architect, in particular, didn't relate to the issues by neglecting or diminishing them, but by pointing out that his team would try to implement them as well as possible at a later point. Nevertheless, the testing results are not directly translated into practice but conflict with it here. The translation of the standard network does not fit with what the architects find relevant and appropriate for their product. This becomes apparent in the discussion of an access configuration issue. After reading out the issue, the architect argued:

"You [the security officer] probably have little detail on how this works in [cloud provider]—there are not the typical IT roles, this is not how we work. Things are more combined, try to differentiate normal developers and specific viewing/billing information".

The officer responds: "But the network security must be ensured".

The architect responds: "This is part of a different process; they are building security groups. Usually, it is whoever manages servers that has access to admins, but this is not possible in [cloud provider], it is too tightly integrated".

The officer is still unsatisfied: "The guys at the data centre do it differently. They set up a server and set a default, for example, someone has access. Hence, the server is not available to others, but only through the [company access control]. Because they can do this, they separate. How do you ensure that it is only accessible by the [internal network]?"

The architect contends: "There are templates. The data centre is a usual part of IT security framework applied at the company. All the things that we deploy are not exposed to the Internet. If they were exposed there would be a route configuration".

Now the officer has a concrete question: "The guys who can configure that aren't the guys who can access it by default?"

Without answering, the architect says: "You cannot solve that by separating the roles. Well, you can do that, but this is not a working situation anymore".

The officer again: "But where is the segregation of duty then?"

The architect: "This is what I am trying to explain, the servers are not exposed to the internet in the first place. They are in a secured network".

It now seems they are talking about two different things: "But the person who sets it up is able to expose it does this without asking anyone".

"Yes, DevOps can do that, but we cannot change that, because otherwise, everything takes longer. It must be realistic. People who are creating servers cannot expose it, except they start writing code. We are segregating duties, but how far do you want to go?"

Now the officer seems to see the light at the end of the tunnel: "Ok, then first you must make a description of that segregation, I couldn't find it anywhere. Then we can discuss it".

Georg wants a word in here: "This is not true. Segregation has grown naturally. We are reviewing it now. A common set of permissions, some of them switch to admin, some to access, some to xyz—and so on. We are reviewing that right now, not everyone needs to be system admin—it will become messy though".

It appears that the rigidity of the assessment that imagines a strict network of controls does not hold up in practice. Instead of aiming at a stable network of relations, in practice, measures were cascaded, swapped out, prioritized, or even dropped entirely as they appear "not suitable" to the system in question.

This scene demonstrates alternative ways of dealing with the difference between the test results and what architects find appropriate. The tension is obvious again as the architect wants to make the specificity of his system clear. The officer replies by suggesting scaling security by referring to the security of the entire network that must be ensured. The architect responds with the specificity of his system. The officer then refers to the standardized way of doing this as it is done in the data center of Powino. Now the architect gets more concrete in his reply, and so the question by the officer is more specific as well (on *"the guy who can configure that"*). But the architect replies this is not a "working solution". Now the officer replies with the standard again: *"segregation of duties"*. The architect replies now to the scaling argument that was used before: segregation of duties replaced (or cascaded) with the exposure to the internet. The officer tries to understand now if the person who sets up the configuration does it alone. He receives a clear answer but also an explanation why this would otherwise not be "realistic". Finally, it is again decided to adjust the documentation to have a better foundation for future discussions.

In this episode, the standard is not simply translated into local action, but neither does the standard nor the technology in development remain untouched by each other. Instead, we hear about ways of *"trying"*, preemptive complying, and later "reviewing". We hear about an ongoing process, some things already being in review right now, and documentation being adjusted to provide a better ground for discussion in the future. This solves a problem that we previously encountered in the regional logic where security testing (i.e., phishing experiments) could have even counterproductive effects when they decreased the trust between employees and the security departments. In this episode, the atmosphere is tense but not hostile. Attempts to increase the reliability of the system do not have the opposite effects but rather are gradually introduced into the system. Although commonly incompliance and incoherence are seen as a problem (as they are in the results of the assessment), the procedure here allow incompliance and compliance to co-exist productively. The episode points to how realistic a controlled network is in digital security; not too much, which does not mean that security mechanisms are set aside.

The topology of flows is better for understanding this form of digital security where even though the network of controls is not entirely rolled out, items are carried forward, dropped, replaced, etc. This perceives digital security as an ongoing process of trying, partially fulfilling, and reviewing how to replace some parts and enter controls in some later shape. This does not cause digital security to completely break down—even the security officer agrees to this—but shows trying to act "as well as possible" with the explicit constraints at hand.

Fulfilling all security definitions as defined by the issues would result in a technology that loses its purpose. On the other hand, security is not completely lost because some test results are not turned into security measures. Other than indicated above, it does not seem like the system breaks down if every issue is not tackled. Some issues are solved prematurely even before the test, and some are not fixed at all. Some issues cause document reviews, others are preemptively introduced, and again others are kept as a reminder in the architect's document folder, possibly to be mobilized later.

It appears that rather than being completely standardized, or completely insecure, the system emerges as fluid; neither depending on enrolling all items nor being completely insecure. Architects try, review, and adjust. The officer follows a trail of security controls that could solve the issue he indicates here. But fixing all the issues would be too "messy" for the system to work. However, security is not completely lost when individual issues are circumvented.

The conversation shows that some testing activities—like negotiating results with developers—engendered digital security as more than a cut-off point, stable indicator, or fixed network. But digital security comes in different shapes and patterns, multiplying with each system that is being tested. Instead of breaking security apart entirely, or abiding by rules completely, the technology remained in a somewhat fragile state, where things are pre-emptively solved, others dropped completely, and yet others may be considered in the future. We cannot assume that the system will ever be "entirely" stable, but may change and evolve further, remaining in a fluid shape. The architects argued for a local adaption of the assessment, suggesting replacing one control with another, and so on. They did not tip over the entire test and throws all the findings overboard; they remained a matter of negotiation, trying, and reviewing. The system is *fluid,* both secure and insecure, in the eyes of the standard. As ANT networks required robustness and strength, it has been acknowledged in post-ANT that some relations are not stable but fluid and therefore robust. Some parts of standardized ways continue while others are adapted or exchanged (Mol & Law, 1994).

In the fluid form of digital security that is practiced through the conversation between the architects and the officer, digital security is not in the form of cut-off points, or as a matter of relations, it gets redistributed and re-configurated by becoming explicit in a particular system. The definition of security is not abstract anymore but entwined with the specific technology. Many options exist for this coming together: "they may be contradictory […] but they may also temporarily become aligned with one another, be partially connected, added together, or even be included in one another" (Denis & Pontille, 2013, p. 13). This coming together is not limited to the moment of the phone call but precedes it: applications were dropped before and will be changed afterward. And the partial connection will continue when some issues remain open. Fluidity unfolds in this mix of standard and local "working situations".

The previous section has introduced descriptions and formulations that pointed to tests that address certain "regions" of digital security, and those that aim at enrolling components into a stable "network". This section goes one step further to investigate how some testing activities include responding to local variance and supporting local adaptation (not without friction) giving shape to a form of digital security as a rather "fluid" composition of standard controls and more. From this section, we have learned that there are some aspects that did not stay in place, controls that were swapped out, and relations that did not stabilize. Other relations may lose their strength over time. Where ANT would see trials of strength establishing the network of relations for digital security, there were moments of negotiation, compromise, and collaboration. In this section, I discuss such adaptation, failing, and compromise as a different pattern of digital security that is performed in the companies. Testing practices in the companies went beyond establishing regions or networks but performed a fluid form of digital security where some things can fall out, be prolonged or subverted. Such tests do not fulfill the characteristics of a trial of strength but instead include adaptation and flexibility.

The section accounts for the fact that presenting digital security as a network leaves out a great deal of uncertainty and ambiguity, on the edges and ends. Employing topological vocabulary has allowed this distinction to be made, and deviation and variance to be particularly included into a model of digital security, something that has so far been considered "deficit" or "incompliant". Engaging with digital security as fluid renders visible the plethora of adaptations and practices that are connected to it without fully realizing it as full conformity. Nevertheless, security stays afloat, things hang together. But, in digital security, this process is not without tension between the developers and the security officer. I want to emphasize the specific precarity that fluid forms of digital security are associated with.

Fluidity and Tension

Consider the ambiguous role of the assessment. In the beginning of the assessment document, it is established that the assessment does not take responsibility for the security of the Data Lake as it does not work as a "full-fledged compliance check of the [name] environment regarding ISO27001".

The officer explains during the phone call that "There will be no final approval for the Data Lake", and that he "just looked at the service, just like a snapshot at things". It is unclear for Georg and the other architect what to do with it then, so the officer adds: "Maybe you can just accept it, accept the risk. There will be no official approval".

To the architect that feels like: "We are playing with words now. You made some recommendations, and we will try to fulfil them. If you are not the final approval, we will consider this as a formal approval".

The security officer adds that the role of security has changed. A couple of years ago, nothing could run without the approval of information security. But this changed and there is no longer a final approval needed.

This seems at odds with the increased importance and investment in security that I have indicated earlier in the chapter. The reason for not giving final approvals is that responsibility for security was increasingly moved into the teams through the appointment of "security champions" or "SecDevOps" teams. This is the consequence of a change in development logics from a waterfall-model process—where technologies are developed aiming at a certain goal—toward more flexible development processes (SCRUM, DevOps), where technologies are developed as new requirements come up and are thus more dynamical and open-ended. As I have previously indicated, the technology in question here is a "*Data Lake*"—a platform that stores heterogeneous data and allows new recombination and analysis. Particularly among data scientists, more flexible work with data was supported, which allows for short-term exploratory and dynamic development. Data science relies on an experimental "trying out" instead of a master plan for development and data analysis. Hence, data analysis and development are less foreseeable and thus cannot be assessed once and for all or fall under the stable character that the standard imagines. The quote from the security officer who sees that "this time" they had to take a different approach rather than certifying the technology before and then letting data in, deciding case by case which date could be uploaded depending on the security measures taken, demonstrates this shift in development processes and the challenges this poses to security. This kind of system continuously produces insecurities and constantly requires adjustments of security. The following conversation between the security officer and the data architect tries to settle the conflict that has been unfolding in the meeting by smoothing the utterances and acknowledging each other's work, which makes them clearly spell out this dilemma.

The ambiguous character of the assessment comes up again toward the end of the meeting when the architect summarizes:

> *"Just to make sure, we are talking about the gold standard here. There should be pragmatism too. Things we want to do, but spent a year on, we cannot do. This is not the right balance. Yes, there needs to be security but otherwise, we will have a very safe system that is as good as no system".*
>
> *The security officer agrees to that and adds: "You are right, nobody wants all the security to be established tomorrow or before data gets analysed, but I would do a bad job if I hadn't pointed out these issues. This is my task to do this. I am not saying you need to do it perfectly. When I do an assessment as an information security officer, and something happens then everyone will blame me. But I think we are on the same page here".*
>
> *To which the architect responds: "I am more than happy for your findings, but by discussing them there should be common sense. There is no value when we work on security for the next year. We need to be clear about what kind of security we want. I fully agree. If you think of something you should mention that".*

Two aspects become very clear from these quotes. First, both parties acknowledge each other's work, thus smoothing the conflict between full compliance or complete negligence. But more than simply being in conflict, the two "sides" show an understanding that different goals and responsibilities are at stake here. Secondly, what becomes clear in the security officer's agreement is that he partly falls back into the logic of regions where responsibility is neatly separated into corporate departments. Fluid digital security is not naïve and feel-good but deals with precarity that is most easily answered with strict (unambiguous and clear) responsibilities. Even though the fluid form is tolerated by the security officer here, it is his formal duty to make clear that it does not follow the standard and thus is hard to account for by him.

This folding into conventional—that is regional—responsibilities is fascinating for fluid digital security, and I will develop it by proposing that fluid digital security is more fragile than the topological literature has so far captured.

Fluid Cybersecurity

To sum up this section, fluid digital security is composed by components translated from a standard as well as shape-shifting local tactics. In the fluid topology, digital security is not performed through conformity—as in a network—or in isolation—as in regions but through flexible, dynamic, and changing local achievements and contestations.

In fluid forms of digital security, "it's not possible to determine identities nice and neatly, once and for all. Or to distinguish inside from outside, this place from somewhere else" (Mol & Law, 1994, p. 660). In fluid forms, components are swapped out, kept in, noted for later, and perhaps partially or entirely dropped. It becomes less easy to say, "this is insecure, or this is secure". In fluid forms, clear cut-off points are missing. Furthermore, it is not easy to determine when the assessment starts and when it ends, showing securing efforts to be ongoing. Lastly, separating between distinct "parts" of the system becomes less attractive when it is seen that controls interlink, overlaps, makes each other obsolete, etc.

Not tipping over, not breaking, but being as secure as possible while also as "realistic" as necessary. Viewing digital security, we find that parts of the system always slip out—in order to become realistic or a "working solution", but they do so without dissociating the entire security. Differing from the metaphor of the network, security is not a single universal standard beyond individual technologies and organizations, but a composite of activities that are held together by relating loosely and constantly changing their form and intensity of relation. Components are exchangeable—if you don't enact security in that way, you can do it in another way. The stability of the infrastructure is not achieved by everything staying in place, but by shifting security controls around. It is stable, workable, and "realistic" because it can remain shapeshifting and fluid.

However, we might conclude that fluidity is somewhat naïve as adaptations are without consequence. I introduce the term "fragility" in order to account for a spatial form that is both fluid, but also precarious;

shape-shifting but about to fall apart. When fragile, fluid patterns are under pressure. In digital security, it must be justified why compliance is not ensured, why the network is left broken, why regions fall out of line, why some controls are switched for others. The term "fragility" accounts for this precarity and tension. Fluidity has not made explicit the precarity and dangers that digital security practices require.

While fluid relations include oscillations and dynamics, when they turn fragile such oscillations have moral connotations; they become a worry and a concern, something to be attentive to. This is not entirely new to fluid forms, but digital security encourages such an emphasis. In fragile forms, what is being tolerated and even essential to uphold fluid forms is put under constant scrutiny.

For the goal of imagining digital security in different patterns, this means it is not enough to try to include local adaptation into the model (nor to see it as local incompliance, which would be a regional model), which encourages thinking with inconclusiveness and discontinuity, but to put more nuance on practices of trying better, being provisionary and in tension. In the fluid pattern of relations, normative good is no longer what follows the standard, but what is locally adapted and often even subverting the standard. But in fragility, this becomes additionally troubling because what is locally good is under constant attack. Fragility can describe when fluid forms become unreliable and troubling.

Making the next step in my argumentation in this chapter is moving from fluid forms that establish robustness through variance and adaptability to asking how such robustness is reached in the case of digital security. Fluid forms of digital security are fragile—they easily tip over. The guiding question for the last section of this chapter is: How is robustness generated? In short, my response is that fragility is answered by the work of doing digital security "as well as possible", carrying along a sublime understanding that those dangers are imminent.

The ambiguity of fluid forms becomes apparent in the last episode where the officer explains it would be his responsibility to point to issues, otherwise he would be blamed. One explanation, mobilizing the terminology of topologies, may be that fluid forms conflict with the regional assignment of responsibility in a company like Powino. Another way of analyzing this tension, however, is to see it as evidence that fluidity, as outlined by post-ANT, is becoming (more) uneasy and conflicting. In digital security, adaptation is not without friction. While cherishing adaptation, one might be inclined to overlook security.

During the phone call, the group switched between requirements of the functionality—the realm of developers—and requirements of security standard—the realm of the security officer—back and forth, forming a third shape that is temporary. The *Data Lake* gains a shifting form, enacted through the phone discussion and negotiations before and after. This switching back and forth, using bits and pieces from the standard and from the existing development, was crucial here. This sees developers and security officers not necessarily on two sides: security on the one side and development requirements on the other but puts the process of shaping the technology in the foreground. I find that this is crucial in how fragile flows are kept together. They rely on tense negotiations and uneasy compromise.

This should not sound like standard security and fluid security are mutually exclusive as if security officers could only think in regions and networks and architects could only do fluid adaptations. On the contrary, it is characteristic of fragile forms that fluidity and stability/rigor overlap and interact. In fluidity, things do not always add up. They are in constant tension. And furthermore, becoming engaged in the complex normativity of shape-shifting digital security is not solely done by one side (officers, developers, users, etc.)

Not Being the "Security Guy"

Standardized security does not align perfectly with local practices. But neither do they exist independently of each other. A fluid shifting of security measures exists—some are fulfilled while others are not. Security officers and architects do not stand on two sides or act as representatives of two different forms of security. Nor is one security better than the other. Instead, we must say that the "good" requires a situated view. Security is "good" in forms of standards but also "good" in operation—depending on the respective practices at stake. I am emphasizing this because this chapter does not suggest throwing all testing out of the window because it performs one form of security or another; rather, it attempts to illustrate how tests generate different patterns of security that aid us in thinking about digital security differently and critically.

Specifically, reducing security officers' work to test and audit makes invisible the various ways in which officers tried to forge good collaborations with developers and employees. Being only "security people", as they were often described by other departments, did not depict them properly. In fact, Stefan, a security officer at Covy had trouble defining himself as a security "guy" and introduced himself with: *"I care about security"* and "That is what I do". This means that he does more than "pushing the rulebook", as he explains to me. Rather he was invested in building rapport with developers, was invited to meetings without asking, and was a constant point of contact for developers. His internal conflict with the discipline of security was that he must do a lot more than follow the standard security in order to "care for security". He tells me that he often sat down with teams of developers and discussed their security issues, realizing that they were entirely different than he had thought. To align their practices with the security controls he had to decide that a standard control (segregation of duty) was not necessary "for now". And he explains that he had rather a "good relationship" with the teams that allows him to make constant comments on procedure than "scare them away". Stefan points out that what he does is care for security, which sometimes requires bending the rules or adapting them to local contexts for a while. Part of his good relationship with developer teams was that he introduced regular code reviews that tested the security of software code. While the code reviews were important, to him it was more crucial that the team would get a positive image of digital security and of him. Rather than checking their performance in the code reviews, he exchanged experiences in the process and addressed problems such as when to ideally perform code reviews. This points to a more comprehensive understanding of testing activities, not only as assurance, but also ways of forging long-term relationships and trusting collaborations. This is something that is hard to test with an abstract performance indicator (such as phishing awareness of standard risk management).

To widen the scope of tests, Siegfried, a security officer at Covy for over 30 years, set up an Excel sheet of risks that people in the company defined for him. The items in the table did not fit official risk categories or risk measurements as indicated by ISO but were filled with situated issues and problems, as he explains to me in a conversation in the cafeteria.

He began with a "smart data risk approach" which he termed his innovatively grounded risk analysis. He tells me that he would talk to people in departments and ask them how they perceived risk in their concrete work tasks. After they had filled out an Excel list of current risks that left a lot of space to state individual risks or non-usual risk perceptions, Siegfried would sit down with the respective people and discuss their list. He would ask "Does that fit?" with what we usually define as risk and "What do we have to change in our risk understanding". Siegfried's risk assessment differed from existing approaches because it combined calculated risks (according to assessment); risks that were tested in audits (e.g., penetration tests); and personal, idiosyncratic risk perceptions.

He found that even listing potential risks already had a cathartic effect: becoming aware of risks enabled people to align their security measures. This was not about security being perfect, but about responding to situated and local risks. "It's not an exact science, it's more like gut feeling", Siegfried explains. But these "gut feeling"-risks are as important as standardized risks as they talk more about the concrete ambiguities of everyday operations. Siegfried tells me that entering an item in the table is followed by a conversation between him and the person who entered it. And often, he admits, entries do not fit the column description. I want to know what he does in such a case. He replies: "Leave it in ". And he explains how the local risk description is more important to him than keeping the table "clean". Inevitably, some risk perceptions also fitted with standardized definitions which remained in the table.

To Siegfried, this flattening of risks from the standard and from "real" practices made the table extremely valuable. In the words of ANT scholars, we can find these sheets to be "qualculative"—both allowing formal calculation and operational judgment (Moser & Law, 2006; Callon & Law, 2005).

The sheet combines methods like "risk maps" where risks are spread out and categorising according to established risk management with information, he gathers in conversations with people all over the company. He tries to comprehend their local risk perceptions—employing both qualitative descriptions and quantitative estimates of risk "values". The challenge to Siegfried was to bring what was reported to him in conversations together with formal reviews, technical vulnerabilities, assessments, and incident reports.

In a slide deck that Siegfried used during a workshop in 2018, he shows me how he tries to cope with "multiple risk views" that he uses as "enrichments" for his table that follows a particular standard ("unique risk format"). Siegfried explained to me showing me the schematic representation of his process, how he classifies between "ad hoc" reactions—under which he subsumes both incidents and "non-compliance" of users—continuous patterns of change like regular "project reviews" and stable ongoing cycles like "criticality assessments".

These three "levels" are similar to what I have described above: Digital security spans across ad-hoc situations, networks, and regions; it is fluid. But Siegfried's approach goes beyond that. The three levels do not remain separate; they must be brought together, and he must go back and forth between them to make his risk assessment. Siegfried explains that his table was updated regularly and thus reminded him to stay flexible in his more standardized risk assessments.

Siegfried's risk approach conveys not only the sheet on which risks are collected but the entire process of sitting down with colleagues and using their experiences in combination with other risks. The result is a complex risk matrix that allows him to better mitigate where there is a need to offer guidance or help. This approach differs vastly from previous ones as Siegfried himself gets involved and engaged. Furthermore, the sheet is flexible and dynamic and allows new categories to be included along the way. In contrast to the approach of regions where measures were calculated according to "cut-off" points or anticipated deficits, this approach engages with and relates between regions (technological measures, organizational measures, etc.). In contrast to the networked approach, risks are perceived to emerge in practice and are subjected to a conversation rather than enrolment/translation.

Siegfried concludes that "*you must allow some liberty in attributes*" and that *"initially concepts and assumptions have held him back"*. Rather than sticking to pre-set categories *"you must tolerate unexpected risk aspects"*. Finally, it is important to him that his table includes "*quality before quantity*". The risk entries must be formulated in a way that others understand them. If this works *"entries in the risk table can cause 'wonders' if you let the experts have the word [...] and only end the discussion when everyone has had the chance"*. Siegfried's risk approach brings calculated and experienced risks onto the same analytic plane.

Sometimes he sends extracts of the table to colleagues, and they begin discussing whether their current security policies still fit with the entries in the table. He must manage to make the table useful for the standardized assessment but also for discussion with colleagues and ensuring that local risk perceptions are still represented. He reports that people even called him when something changed, actively supporting his approach (whereas it was usually perceived that security would "come in" and disturb things).

Siegfried's sheet embodies views on risk that are not steeped in security terminology but are based in practice. Nevertheless, they are an object of clustering and calculating a risk score. While the score always stays open for reconfiguration, it is necessary to link local risk perceptions to formal assessments and quantitative risk calculations.

Siegfried's table is an example of digital security that is both fluid—adapting, changing, and in flux—but also stable and reliable: the table allows the fluid to fit into the standard and enable the conversion between the local and the standardized. The pattern of relations in which Siegfried enacts digital security with his table is a tense friction between fluid and rigor. He goes back and forth every time he opens the table for new entries, and something must be put into a calculative form when reaching out to the standardized assessment. In fragile forms, specific efforts exist to make fluid forms continuous.

Siegfried's table was not a one-off, it did not stick together smoothly, but required continuous attention and adaptation work. The entering of items and adjustment of categories needed to be done again and again, in each new project, for each new assessment and conversation.

As with the security officer and the developers in the assessment above, switching between fluidity and other topologies causes tension and ruptures. But it is in these ruptures that much of digital security work is happening, not abiding by nor completely detached of formality, but also not "only" fluid. If anyone let go—either the developers doing fluid adaptation, the security officers approaching networks or searching for regions—digital security would break down. Digital security dwells in this frailness and fragility—a constant process of locating, networking, fluid adaptations, constant back and forth, and negotiating. Dealing with this fragility means continually adapting, trying to make regulations, formal responsibility, and local adaptation work together. The different versions of digital security enacted by the tests (as regions, network or fluid) may be completely separate from one another, while in other situations, they may temporarily align, be partially connected or conflict.

Implications 1 and 2

Acting in fragile computing, this chapter has argued, does not stay in one topology. Neither topology is superior or beneficial. There are good reasons to approach digital security as a matter of regions: islands of incompliance or spaces of unawareness. But and that is the point I hope to make here, there are other ways of understanding digital security and thus other means of acting upon it. With the specification of a fragile fluidity, I suggest that while fluid digital security is an as valid pattern as regions or networks, it is also more conflicted and uneasy in this field than previously thought in ANT.

Over the course of my research, I have come to appreciate the work of developers and security officers alike, especially after seeing the efforts, practical and moral ambiguities, and tensions they were often caught up in. I have come to term it as their capacity of enacting fragility: switching between regions, networks, and fluidity. They help us to understand a different form of digital security—not only compliance and standards—but practices of tinkering, consensus-seeking, and negotiation. Security officers like Siegfried are opposed to attempts to scare employees into complying with security regulations and is engaged in careful adapting of regulation to local needs and vice versa. Mark Burdon and Lizzie Coles-Kemp (2019) observe that the "role of the information security practitioner shifted from the techie, the regulator or the security auditor to the articulator, the facilitator and the messenger of information security" (p. 13). I want to add that occasionally they must be a negotiator and step away from making abstract and normative claims to engage with local troubles. Burdon and Coles-Kemp argue further that "[security officers] adapted their role and purpose to negotiate the conflict and complexity of their environment. In this environment, information security was not about controls and standard compliance" (ibid.), it becomes a more tacit mode of knowing security that is situated within the practices. But as I have noted before, it is not useful to see the work of keeping fragile forms together solely at the hand of security officers. Dealing with fragility can also be found in other practices.

Fragile Form

Fluidity cannot fully grasp what is going on in adaptations of digital security. Flows lack the pressure and prevalence of calls for accountability that are part of digital security. I call this specific characteristic of fluid spaces in digital security "fragile", as more than previous considerations, it carries volatility and vulnerability of fluid actions. In the fragile form, practice cannot be reduced to "tinkering" without consequence; instead, moral accountability and responsibility are configured.

We need a way of thinking about, working with and within systems that are not only incoherent and fluid, but that are fragile—prone to breaking; locally subverting, unruly and can be dangerous. Law (2000) notes in his analysis of a train incident with several deaths that the tendency after such an incident is to aim for repair and becoming better in predicting and preventing such disasters (p. 14). This is of course important. But, he states, something is missing: Thinking about a system's imperfections includes more than locating sloppiness, finding imperfection to be a curse to overcome, to fix. He acknowledges that this is particularly hard to suggest when people have died and societies want reliant infrastructures, but it is important to acknowledge that stable or not (compliant or not, secure or not, etc.) are no opposites.

I have argued in this chapter in a similar way of thinking about digital security incompliance not as a curse but as an opportunity to understand how systems hold together despite and under specific fragility. Digital security entails formal organization accountability, which causes a specific frailness, and uneasiness that is the reason for conflict in the companies. Fragility captures this precarious state of fluidity—where fluidity is always considered a threat to networks and regions because it allows variants that *might* turn out to be dramatic. Digital security shows the precariousness that fluid topology does not entail. Digital security renders technologies delicate, speculative, susceptible, and exposed.

Fragile forms resemble what Law and Mol introduce as "fire" (Law & Mol, 2001): a flickering and shape-shifting spatial form that exists in between presence and absence, what can be and is, what should be and shouldn't. Less articulated, however, is that fire is risky, can act out, or overburden (Law & Singleton, 2005). Fragility can emphasize this: fragile forms are flickering, risky, and frail at the same time. In fragility, stability is reached not only by being simply shape-shifting but also by imaging what could be, by oscillating between control and loss of control, by being secure and insecure. Fragility is more precarious than just fluid. Perceiving of digital security as fragile allows us to stay with the intricate entanglements, inconclusiveness, and productive tensions central to digital security work in organizations. The third type of testing activities—those that account for fluidity and even more for fragility—are important to organizations but are often neglected when security is perceived as a choice, rational decision making, or a simple translation of standards. It emphasizes tense negotiation, trust, and collaboration.

Security experts would probably agree that "perfect or total security is an illusion" (Tøndel et al., 2020). As Tøndel et al. argue, this imperfection is even foundational for risk management approaches as outlined by standard framework such as ISO 27001. Instead of assuming perfect security, they argue for shifting attention toward how it is determined in practice as to what "good enough" security means. Something that can hardly be determined from afar, "good enough" security includes "trade-offs, not absolutes" (p. 1) and a shifting assessment of objective digital security. This fits my analysis of a fragile form of digital security that relies on in-situ adaptations, yet nonetheless a precarity that is carried along.

I want to point to some directions such work could take: In *Arts of Living on a Damaged Planet*, Tsing and colleagues suggest inquiring into practices that are related to the possibility of end and fragility than continuing with dreams of stability and progress. Digital security is a form of attending to infrastructures that can no longer be "patched up" into a stable network, but that require attention to be paid to ambiguous and imperfect forms. In Tsing's terms, living in "ruins" means not only dealing with what is given but also "stimulates noticing, as one works with what is available" (Tsing, 2015, p. 278). Such noticing is open to ambivalence and fluidity but does not assume things to be endless and ongoing. Living with precarity, Tsing notes "requires more than railing at those who put us here [...] We might look around to notice this strange world, and we might stretch our imaginations to grasp its contours" (ibid., p. 3).

In her example of a mushroom species that thrives in industrial forest areas, Tsing's argument is not to "go back" to a "natural" state, but to find ways of living with and enduring the past without rushing to solutions. In such times, single entities are not easy to pick out and fix, neither does it suffice to speak of a robust network. No innovation will bring us beyond the destruction we have caused, and no innovation will finally fix insecure technologies. But what we can see are practices of attending to, caring for, and enduring insecurity that are worth investigating. Precarity is the keyword here—both for Tsing and for my concept of digital security. Fragility accounts for the work of complex moral registers and formal accountability in fluid forms of improvisation and local adaptation. It does not neglect the hard-to-make decisions and modes of taking over responsibility for wide-ranging sociotechnical worlds but makes them part of the practices of attending to digital security. Digital security cannot be all improvisation and fluidity, nor can it be all rules. Like dealing with environmental damage, it needs responsibility as much as variance and hope.

The STS definition of disaster foregrounds the failure of a diverse and nested system that cannot be confined in time or space, nor adequately addressed with standard procedures and modes of thought (Fortun et al., 2017 p. 1004). Disasters upset long-standing social theoretical assumptions that systems are functional, with components working autonomously but interactively to produce stability. Shifting attention to dysfunction thus includes a shift in theoretical repertoires, one that includes instability, dysfunction, and fragility. STS approaches to natural disasters, industrial spills, and pollution have mapped out possible paths; however, they have not included technological disasters and digital security incidents in their theorizing.

What is important here is that they witness a shift from those patterns of relations that rest upon stability and continuity (even if enabled by adaptation and fluidity), toward patterns that include disasters and possible endings. Fragility is not endless; on the contrary, it is precarious and susceptible.

In a fragile pattern, disruption and discontinuity are more immanent than in fluidity, networks, or regions. While fluidity accounts for continuity that is enabled by adaptation and flexibility—instead of conformity—fragility accounts for spatial forms that are inherently broken and always on the edge of falling apart. It is not that overly abrupt changes will threaten the continuity, as it is with fluidity, but that it is fragile forms that are dancing on the knife edge and volcano's ridge. This dance combines adaptation and engaging with standards and networks. It combines regions with fluids and networks, which makes it possible but still inherently volatile.

While illustrating three forms would have sufficed to give readers an idea of how digital security in companies is not only a matter of regions, which is most prominent in academia, but also of networks and flows; while spending time in the companies, my fascination grew regarding everyone's capacity to make a compromise, and to bend the rules while being morally affected by what is "good" security. To account for this tension and friction, I have introduced the specification "fragile" for digital security.

Fragile shares its reference point with fluidity in varying and dynamic forms. But what is fragile is more conflicting, more controversial. At its core lies a "good" that is unreachable. It has a dimension of moral authority and a higher stance that is hard if not impossible to reach in practice. But actors persevere, try to come up with relevant ways, and include creative acts in their endeavors of securing, all the while being aware that things are fragile. Security officers have realized the need for adaptations and compromise, despite (and in tension with) being loyal (and responsible) to standards. The two are not separate; they conflict, align temporarily, but never fully reach a conclusion.

This chapter has talked about digital security without explicitly defining it beforehand while developing different forms from the empirical material. This is to leave the process open to see what digital security is practically constituted of instead of settling on its definition a priori. By not defining digital security beforehand we were able to wander around for a bit and see what it entails and what differences we usually overlook when simply speaking of digital security as a homogeneous set of activities that counteract attacks. I have put forward an understanding of digital security as formed in different topologies. It must be emphasized that topologies co-exist—in fact, they take different lenses and in turn are capable of responding to different facets, and possibly solving different problems. Having said that, it is noteworthy that digital security is most commonly enacted as "regions": differentiating between humans and technologies, places in which digital security is better than in others, and so on. Seeing digital security as a matter of regions, we can ask different questions, such as: How well designed are digital security measures? How capable are employers of identifying phishing mail? How much economic effort does it take to improve digital security? And is it worth it? These are highly relevant inquiries, but not the only ones, as this chapter argues.

Pondering about networks is another way of inquiring digital security. Questions here would be: How does an incident come into being across more-than-technological networks? (Stachel & DeLaHaye, 2015). How successfully are actors enrolled and how well are their interests aligned? (Hedström et al., 2010). How does securing mobilize various networks and concerns?

Yet another approach is tailored toward practice. It is not enough to trace "errors" in the network of human and nonhuman actors that uphold digital security. Rather it is important to shift attention to practices where fragility is neither put aside nor completely stabilized. What is assembled in digital security, and how it is achieved or partially achieved? Such questions are at the center of efforts in the companies concerned with forging long-term trust and a ground on which conversations can maintain and negotiations continue.

This applies the lens of some contemporary theories in anthropology and STS on the irreversibility of damages to the planet to matters of digital security. Computational systems show comparable fragility: there are irreversible dangers in the close entanglements of devices and critical services, but there is no easy solution to overcome that. We must learn to live with the tension introduced through fragile forms, the potential drama and risk they involve. But not in the sense of limiting the unknown through statistical extrapolation, but rather through investigating the pragmatic ways actors deal with fragility. In the topological vocabulary, this encourages theorizing local variance and precarity together. Besides trials of strength and explicitness, the work of security officers and developers shows trails of uncertainty—how much can we tolerate leaving out? Under what circumstances is exchanging one control for another tolerable? What ambiguity can we live with? For how long?

What this introduces is a method to trace how digital security is enacted in testing practices. This is not to say that one form is better than another, but simply to acknowledge that different forms of tests, and different patterns of relation of digital security exist. Some are more suitable to formal ways of accounting, e.g., cut-off points and stable networks, whereas others are rather hard to generalize. Digital security facilitates an in-between version.

Implication 1: A Taxonomy of Digital Security Testing

Rather than capturing tests and assessments as tailored toward threats, regulations, or existing measures as done in existing typologies, the chapter has suggested conceiving of testing activities as creating patterns of relations.

As indicated above, testing is never all-encompassing, but creates specific inquiries into the world, while leaving others aside (Strathern, 1991). This points to the power of testing in allocating resources and responsibility and creating dominant notions of objects in question. The tests in this chapter have differed depending on how they dealt with ambiguity and with what falls through the grids of their formal accounting. It is important to state that *testing practices* (rather than tests) do include significant amounts of ambiguity and uncertainty, something that only frequently comes through to the results, but becomes visible when looking at accompanying conversations, negotiations and the planning and employment of tests. When looking at such practices, testing does more than produce (partial) views of digital security; it allows for variance and fragility to emerge and be endured.

Testing incorporates understanding local requirements, sitting down for conversations, finding compromises, exploring options, and forging long-term relations and trust. It needs the willingness to understand and accept the practices and risk perceptions of developers and users instead of stable assurance. Research into the performativity of digital security tests (across all topologies) is needed to investigate what they exclude and include, and how regimes of visibility and invisibility affect the ways digital security is imagined and allocated. Furthermore, what I have emphasized in particular in this chapter is to build a better understanding of how testing allocates responsibility for digital security, whereas regional forms have been widely dominant. Employing the topological model suggested here may shed light on other possibilities of framing digital security responsibility.

The chapter develops current efforts in digital security research to improve assessments and tests further by pointing to necessary blind spots and adaptations that testing practices include. Rather than providing an unambiguous picture of digital security, testing practices show negotiation, compromise, and possibility for local adaptation; but they also problematize and bridge formal accountability and local ambiguity.

Because digital security remains notoriously inconclusive, dynamic, and fragile, digital security testing requires more than "checking" boxes; it needs the work of committed security officers who can translate digital security regulations into concrete and specific tasks for developers. Such translation work was extensive at Covy and Powino and requires more scholarly attention (Williams & Coles-Kemp, 2019). Testing digital security is not solely about producing evidence and lists of "problems" (or fixes), but it engages in and is responsive to compromise and adaptation. Perceiving the work of security officers and developers in finding and sustaining compromises and fragility points us to the material work of digital security much more than pretending stability and judging or blaming incompliance. Fragility appeals to staying with what is the tough work of digital security, which is hard to make decisions, and ambiguities that are tolerated and carried forward in practice. It foregrounds the gray areas of digital security work.

Further questions include: How can testing better incorporate fluidity and fragility? How can tests become devices to curate ground for negotiation and trust between developers and security officers? How can testing activities be designed to incorporate diverging models of responsibility?

Implication 2: A New Spatial Form to Think with

The chapter has argued that digital security is most commonly approached in testing in the form of a regional pattern or network. What some testing practices point to is that there is also uncertainty and ambivalence that is characteristic to digital security. This is hard to account for when framing the goal of security work as compliance and assurance. More often, security officers and developer act in gray areas.

I have explored topological literature to identify different ways of attaining robustness in digital security that incorporates variance and ambiguity. Whereas STS scholars have argued that flexibility and stability are not opposites, the concrete ways in which things are held together despite fragility are not fully conceptualized (nor will they ever be).[5] Existing topological approaches have often attended to technological systems (De Laet & Mol, 2000; Sørensen, 2009); or semi-standardized procedures, such as medical diagnosis (Mol & Law, 1994); or interdisciplinary collaboration (Star & Griesemer, 1989); but have not attended to digital security matters. This poses some challenges I have developed in this chapter.

---

[5] I make use of the term fragility even though—being akin to fluidity—it is hard to speak of as a noun. More correctly, it would be employed as an adjective, which however suggests there was anything more stable to reference it too, e.g., "fragile technology". Keep in mind that saying both "fluidity" and "fragility" as nouns does not fully capture the fact that something being fluid/fragile is never easily one thing. Others have run into this problem, for example, Karen Barad who decided for "intra-action" as a way of perceiving of constant co-creation; or Haraway who has employed the term "hybrid" (yet not "hybridity"). To improve accessibility, I will sometimes say "fragility" meaning that which is constantly changing its fragile shape.

In the form of *fragile flows*,[6] fluidity is accompanied by both moral and judicial responsibility that are distributed alongside objects. Fragile spaces entail friction: friction between digital security and developers, where local variance is not without tension, but tolerable amounts of variance must be negotiated and discussed.

Fragility is not open-ended. When considering fluid forms as fragile, we pose questions about their potential breakdown; how to handle their variance in the face of imminent danger.

This echoes the realization that technologies been suggested as the fix for social problems and seemingly endlessly fixable, become problematic and no longer repairable. They are "ruins" in Anna Tsing's terms, where it is important to perceive practices as ruins, rather than thinking of how to overcome them. Dealing with fragile relations involves a fascination and appreciation of fragility and breakdown, which counteracts a progress bias in science and STS (Jackson, 2014). Attending to fragility in the case of digital security brings us closer to stories of maintenance and care that transgress and become partially visible as strict orders of modernity come apart (Puig de La Bellacasa, 2017). Such an interest in the fragility of things barely holding together is contemporary, and important in anthropology that is confronted with anthropogenic disasters (Gan & Tsing, 2018). However, digital security shifts attention somewhat to the ruins of digital infrastructures and modes of living with digital (in)security. How then does applying the increased sensitivity in STS of more-than-human relations and fragile upholding change its scope? The notion of fragility developed in this thesis conveys such a reach.

The lens (or "trick") of topologies has attempted to analyze more closely how digital security is assumed to attain robustness. The topology

[6] The addition of a new spatial metaphor has the tendency to simplify previous spatial forms and suggest that "old" ones are fully left behind (Sørensen, 2009). The same is the case here. Fluidity is reduced to what is constantly changing, which does not necessarily include the urgency and drama that fragility does. But there is drama in fluidity as well (e.g., taking responsibility for machines and people in De Laet & Mol, 2000). Still, this offers a conceptual contribution with which to further develop topologies. I have debated with many whether fragility adds a spatial form or is a re-emphasis on the precarity and risk of fluidity, more prominently now as the world seems "on the edge". Fluidity is not left behind—the patterns do not relate regionally (they do not form a neat mosaic). Law and Singleton (2005) discuss "fire" as an additional shape within the register of fluidity that allows to conceptualize oscillating presence and absence. Fire is fragile: dangerous, a source of risk as well as precious and delicate.

lens allows a taxonomy of tests according to how they address security and which aspects they fail to address. This is important for how we assume digital security is assured by testing. But more than offering a taxonomy of tests in practice, the topology lens has aimed to think about compromise and variance as crucial parts of digital security rather than as something that can be neglected and is largely invisible. Robustness through variance—rather than through sameness or distance—is fundamental to the fluid pattern and something that has been emphasized in this chapter. Nevertheless, I did not stop there because the tensions and frictions of digital security variance must also be accounted for. In particular, I argue, digital security demonstrates that alterations must be responding to control and fixity. I find this worth noting and have argued that it needs a slight addition to the pattern of fluid relations that entails this precarity. Thus, I suggest the notion of fragility. What is fragile is prone to failure; it needs forms of action sensitive to its precarity. Fragile fluidity cannot be simply eradicated; variance and failure are part of it. It needs to stay with the trouble, with situations that are uneasy but should not be avoided (Haraway, 2016).

What does dealing with fragility mean in practice? What work is necessary to maintain stability in friction and in spite of fragility? The next chapter develops this question. In fragility, digital security is attended to as a matter of ongoing and constant attachment, conflicting imaginaries, and ambiguous decision making. It is acted upon in moral gray-areas. Instead of making rational decisions of secure or insecure action, enacting digital security is often inconclusive and what is good and bad only emerges from concrete action. In short, the next chapter argues digital security is enacted as a matter of care.

References

Akrich, M. (1992). The de-scription of technical objects. In: W. E. Bijker und J. Law (Hg.) *Shaping technology. Studies in sociotechnical change*. MIT Press (Inside Technology), 205–224.

Al-Matari, O. M., Helal, I. M. A., Mazen, S. A., & Elhennawy, S. (2018). Cybersecurity tools for IS auditing. In *2018 Sixth international conference on enterprise systems (ES)* (pp. 1–6). IEEE.

Bada, M., Sasse, A. M., & Nurse, J. R. C. (2019). *Cyber Security Awareness Campaigns: Why do they fail to change behaviour?* http://arxiv.org/pdf/1901.02672v1

Barry, A., & Born, G. (2013). Interdisciplinarity: Reconfigurations of the social and natural sciences. In A. Barry & G. Born (Eds.), *Interdisciplinarity: Reconfigurations of the social and natural sciences* (pp. 1–56). Routledge.

Berg, M., & Timmermans, S. (2000). Orders and their others: On the constitution of universalities in medical work. *Configurations, 8*, 31–61. https://doi.org/10.1353/con.2000.0001

Bowker, G. C., & Star, S. L. (2000). Sorting things out: Classification and its consequences. https://doi.org/10.7551/mitpress/6352.001.0001

Bravo-Lillo, C., Cranor, L. F., Downs, J., & Komanduri, S. (2011). Bridging the gap in computer security warnings: A mental model approach. *IEEE Security and Privacy, 9*(2), 18–26. https://doi.org/10.1109/MSP.2010.198

Burdon, M., & Coles-Kemp, L. (2019). The significance of securing as a critical component of information security: An Australian narrative. *Computers & Security, 87*, S. 101601. https://doi.org/10.1016/j.cose.2019.101601

Callon, M. (1984). Some elements of a sociology of translation: Domestication of the scallops and the fishermen of St Brieuc Bay. *The Sociological Review, 32*(1_suppl), 196–233.

Callon, M., & Law, J. (2005). On qualculation, agency, and otherness. *In: Environ Plan D, 23*(5), 717–733. https://doi.org/10.1068/d343t

De Laet, M., & Mol, A. (2000). The Zimbabwe bush pump. *Social Studies of Science, 30*(2), 225–263. https://doi.org/10.1177/030631200030002002

Denis, J. & Pontille, D. (2013). *Material ordering and the care of things*, CSI Working Papers Series 034, Centre de Sociologie de l'Innovation (CSI), Mines ParisTech. https://ideas.repec.org/p/emn/wpaper/034.html.

Donaldson, S. E., Siegel, S. G., Williams, C. K., & Aslam, A. (2015). *Enterprise cybersecurity. How to build a successful cyberdefense program against advanced threats*. Apress Springer Science+Business Media.

Dourish, P., Grinter, R. E., La Delgado Flor, J., & Joseph, M. (2004). Security in the wild: User strategies for managing security as an everyday, practical problem. *Personal and Ubiquitous Computing, 8*(6), 391–401. https://doi.org/10.1007/s00779-004-0308-5

Fortun, M., Fortun, K. & Marcus, G. E. (2017). Computers in/and anthropology: The poetics and politics of digitization. In Larissa H.; Horst, H. A.; Galloway, A. & Bell, G. (Eds.) *The Routledge companion to digital ethnography* (pp. 37–46). Routledge Taylor & Francis Group. https://www.taylorfrancis.

com/chapters/edit/10.4324/9781315673974-11/computers-anthropology-poetics-politics-digitization-mike-fortun-kim-fortun-george-marcus

Gan, E., & Tsing, A. (2018). How things hold. *Social Analysis, 62*(4), 102–145. https://doi.org/10.3167/sa.2018.620406

Hanseth, O., Monteiro, E., & Hatling, M. (1996). Developing information infrastructure: The tension between standardization and flexibility. *Science, Technology, & Human Values, 21*(4), 407–426. https://doi.org/10.1177/016224399602100402

Haraway, D. J. (2016). *Staying with the trouble*. Duke University Press.

Hedström, K., Dhillon, G., & Karlsson, F. (2010). Using actor network theory to understand information security management. In K. Rannenberg (Ed.), *Security and privacy - Silver linings in the cloud: 25th IFIP TC-11 international information security conference, SEC 2010*, Held as Part of WCC 2010, Brisbane, September 20–23, 2010 (pp. 43–54).

Hutchins, E. (1996). *Cognition in the wild*. Bradford Books.

Jackson, S. J. (2014). 11 Rethinking repair. In Gillespie, T., Boczkowksi, P. & Foot, K. A. (Eds.) *Media technologies: Essays on communication, materiality, and society* (pp. 221–239). https://doi.org/10.7551/mitpress/9042.003.0015.

Klimburg-Witjes, N., & Wentland, A. (2021). Hacking humans? Social engineering and the construction of the "deficient user" in cybersecurity discourses. *Science, Technology, and Human Values, 46*(6), 1316–1339. https://doi.org/10.1177/0162243921992844

Kreitz, M. (2020). Security by design in software engineering. *SIGSOFT Software Engineering Notes, 44*(3), 23. https://doi.org/10.1145/3356773.3356798

Lampland, M. (Ed.). (2009). *Standards and their stories. How quantifying classifying and formalizing practices shape everyday life*. Cornell University Press.

Latour, B. (1990). Technology is society made durable. *The Sociological Review, 38*(S1), 103–131.

Latour, B. (1996). *Aramis or the love of technology*. Harvard University Press.

Latour, B. (1999). *Pandora's hope. Essays on the reality of science studies*. Harvard University Press.

Latour, B., & Venn, C. (2002). Morality and technology. *Theory, Culture and Society, 19*(5–6), 247–260. https://doi.org/10.1177/026327602761899246

Latour, B., Jensen, P., Venturini, T., Grauwin, S., & Boullier, D. (2014). The whole is always smaller than its parts'. *A Digital Test of Gabriel Tarde's Monads*, 31.07.2014. https://hal-sciencespo.archives-ouvertes.fr/hal-01053567/document

Law, J. (1987). On the social explanation of technical change: The case of the Portuguese maritime expansion. *Technology and Culture, 28*(2), S. 227. https://doi.org/10.2307/3105566

Law, J. (1999). After ant: Complexity, naming and topology. *The Sociological Review, 47*(1_suppl), 1–14. https://doi.org/10.1111/j.1467-954X.1999.tb03479.x

Law, J. (2000). *Ladbroke Grove, or how to think about failing systems*. http://www.comp.lancs.ac.uk/sociology/papers/Law-Ladbroke-Grove-Failing-Systems.pdf

Law, J. (2002). *Aircraft stories. Decentering the object in technoscience*. Duke University Press (Science and cultural theory). http://site.ebrary.com/lib/all-titles/docDetail.action?docID=10207714

Law, J., & Mol, A. (2001). Situating technoscience: An inquiry into spatialities. *Environment and Planning D: Society and Space, 19*(5), 609–621. https://doi.org/10.1068/d243t

Law, J., & Singleton, V. (2005). Object lessons. *Organization, 12*(3), 331–355. https://doi.org/10.1177/1350508405051270

Mol, A. (2003). *The body multiple: Ontology in medical practice*. Duke University Press. https://doi.org/10.1215/9780822384151

Mol, A. (2008). *The logic of care*. Routledge.

Mol, A., & Law, J. (1994). Regions, networks and fluids: Anaemia and social topology. *Social Studies of Science, 24*(4), 641–671. https://doi.org/10.1177/030631279402400402

Moser, I., & Law, J. (2006). Fluids or flows? Information and qualculation in medical practice. *Info Technology and People, 19*(1), 55–73. https://doi.org/10.1108/09593840610649961

Muniesa, F., & Linhardt, D. (2011). Trials of explicitness in the implementation of public management reform. *Critical Perspectives on Accounting, 22*(6), 550–566. https://doi.org/10.1016/j.cpa.2011.06.003

Puig de La Bellacasa, M. (2017). *Matters of care: Speculative ethics in more than human worlds*. University of Minnesota Press. (Posthumanities, v.41).

Rahim, N. H. A., Hamid, S., Mat Kiah, M. L., Shamshirband, S., & Furnell, S. (2015). A systematic review of approaches to assessing cybersecurity awareness. *Kybernetes, 44*(4), 606–622. https://doi.org/10.1108/K-12-2014-0283

Sabillon, R., Serra-Ruiz, J., Cavaller, V., & Cano, J. (2017). A comprehensive cybersecurity audit model to improve cybersecurity assurance: The CyberSecurity Audit Model (CSAM). In *Proceedings of the 2017 international conference on information systems and computer science (INCISCOS 2017)* (pp. 253–259). IEEE.

Sørensen, E. (2009). *The materiality of learning: Technology and knowledge in educational practice (Learning in doing: Social, cognitive and computational perspectives).* Cambridge University Press. https://doi.org/10.1017/CBO9780511576362

Stachel, R. D., & DeLaHaye, M. (2015). Security breaches in healthcare data: An application of the actor-network theory. *Issues in Information Systems, 16*(II), 185–194.

Star, S. L., & Griesemer, J. R. (1989). Institutional ecology, 'translations' and boundary objects: Amateurs and professionals in Berkeley's museum of vertebrate zoology, 1907–39. *Social Studies of Science, 19*(3), 387–420. https://doi.org/10.1177/030631289019003001

Strathern, M. (1991). *Partial connections.* Rowman & Littlefield (ASAO special publications, 3).

Strathern, M. (2000). The tyranny of transparency. *British Educational Research Journal, 26*(3), 309–321. https://doi.org/10.1080/713651562

Suchman, L. (2006). Human-machine reconfigurations: Plans and situated actions. In *Learning in doing: Social, cognitive and computational perspectives* (2nd ed.). Cambridge University Press. https://doi.org/10.1017/CBO9780511808418

Timmermans, S., & Berg, M. (1997). Standardization in action: Achieving local universality through medical protocols. *Social Studies of Science, 27*(2), 273–305. https://doi.org/10.1177/030631297027002003

Tirumala, S. S., Sarrafzadeh, A., & Pang, P. (2016). A survey on internet usage and cybersecurity awareness in students. In *2016 14th annual conference on privacy, security and trust (PST)* (pp. 223–228). IEEE. https://doi.org/10.1109/PST.2016.7906950

Tøndel, I. A., Cruzes, D. S., & Jaatun, M. G. (2020). Using situational and narrative analysis for investigating the messiness of software security. In *Proceedings of the 14th ACM. ESEM '20: ACM/IEEE international symposium on empirical software engineering and measurement* (pp. 1–6). ACM Digital Library. DOI: https://doi.org/10.1145/3382494.3410701.

Tsing, A. L. (2015). *The mushroom at the end of the world: On the possibility of life in capitalist ruins.* Princeton University Press.

Williams, P. A. H. & Coles-Kemp, L. (2019). Teaching as a collaborative practice: Reframing security practitioners as navigators. In Zhigeng Pan, A. D., Müller, C.W., Mingmin Zhang, A. E. R., & Kifayat, K. (Eds.) *Transactions on edutainment XV, Bd. 11345.* Springer (Springer eBooks Computer Science, 11345) (pp. 108–128).

3

Tinkering

This chapter centers on practices of tinkering with access policy code. Access policy is a critical security function in most systems, determining what rights individual users, or groups of users, have (right to edit, to delete, etc.). In this chapter, two developers experiment with the effects of access policy code by making various changes in the code text. Their engagement with digital security unfolds as tinkering and experiencing its outcome. During the process they articulate their own 'disconcertments' (Verran, 2001), moments in which they are confronted with formal security guidelines and their practical and situated needs. The chapter argues digital security practices like these can be understood as care: material work that has ethico-political commitments. The chapter argues that the ethics of digital security—or questions surrounding what "good" or "bad" security is—are not finally settled in practice. Rather than a matter of abstract rule, 'good' digital security rests in practice. The chapter therefore suggests to understand the morally conflicting and uneasy work of digital security.

The previous chapter has attended to digital security in the form of testing. It has been argued that testing is not always aimed at producing robust knowledge to evaluate the security of a system, but also involves developing local compromise, as well as impermanent and fluid forms of security. The following empirical story shows Karolin and Georg as they write,

L. Kocksch, *Fragile Computing*, https://doi.org/10.1007/978-981-99-9807-4_3

copy, and test access policy code of the Data Lake technology that was subject to a formal assessment in a previous chapter. Karolin and Georg negotiate between each other and bits of code as to how much access Karolin should get to the new data system. Access control policy code is part of the authentication and authorization mechanisms of a system. Such mechanisms determine what group of users gets access to what parts of a system and specify what kind of rights are attributed. Rights differ from being able to read-only, edit, delete, or give permissions.

Setting Access Control

Access to data is controlled by establishing authentication mechanisms (passwords, PIN codes, multiple factors) and authorization policies (what actions an authenticated user can perform). Figure 3.1 shows the "Allow" command's specifications: "Allow" "Iam: Get" or "Iam: List". "Iam" defines the user group that is configured separately. It determines which users belong to which user group, e.g., admin, users, external, internal, etc., who are then allocated permissions in the process described here. The "list" command allows a specific user group to list files in a folder. In Fig. 3.2, more "Actions" are allowed, e.g., the command "DeleteUserPolicies", which allows a user to delete other users'

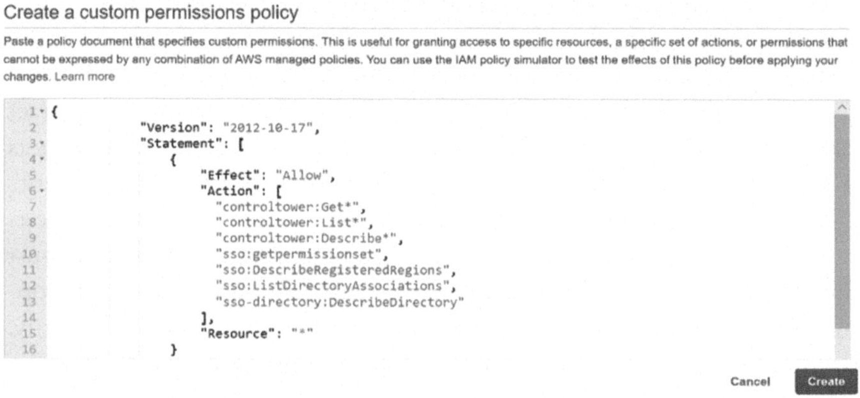

Fig. 3.1 Access policy code (https://aws.amazon.com/de/blogs/security/how-to-get-read-only-visibility-into-aws-control-tower-console/ by Bruno Mendez 2020)

```
{
    "Version": "2012-10-17",
    "Statement": [
        {
            "Sid": "AllowUsersToPerformUserActions",
            "Effect": "Allow",
            "Action": [
                "iam:ListPolicies",
                "iam:GetPolicy",
                "iam:UpdateUser",
                "iam:AttachUserPolicy",
                "iam:ListEntitiesForPolicy",
                "iam:DeleteUserPolicy",
                "iam:DeleteUser",
                "iam:ListUserPolicies",
                "iam:CreateUser",
                "iam:RemoveUserFromGroup",
                "iam:AddUserToGroup",
                "iam:GetUserPolicy",
                "iam:ListGroupsForUser",
                "iam:PutUserPolicy",
                "iam:ListAttachedUserPolicies",
                "iam:ListUsers",
                "iam:GetUser",
                "iam:DetachUserPolicy"
            ],
            "Resource": "*"
```

Fig. 3.2 Access policy code identity management AWS

permissions. Each of the green lines specifies an action that is allowed for users of this group. If access control code is written well, security officers would argue, every line would be adapted dynamically depending on changing constellations and needs of the user group.

Two varieties of setting policy code exist: Either explicitly stating what areas a user group gets access to (a positive definition), or explicitly stating what areas a user group does not get access to (a negative definition). The former is considered more secure because access is not by default given and then revoked. The code in this episode follows the positive definition model. This means that instead of revoking certain tasks—and leaving potentially unknown or contradictory rights unset and thereby allowing them—this process allows certain tasks while all others remain forbidden. In that way, it is assumed, policies do not get as easily overwritten by more general policies. However, putting an asterisk ("*") in the "Action" code, for example, means that all actions are allowed. An asterisk thus partially subverts the minute allocation of rights by acting as a blank. This would be discouraged by security officers as it sets vaguely defined access.

It is summer outside, and I am sitting in an air-conditioned office with large windows with a view over the city centre. I am in the office of Georg and Karolin, two developers at Powino's newly found data analytics team. Georg calls himself the "junior Data Lake architect", a term he made up for himself to indicate the new responsibilities that came along with the new technology. As such, he is responsible for granting others permission to a cloud service, the centrepiece of the company's new cloud architecture. Karolin is a developer who had taken over several projects over the last two weeks that should use the cloud to store their data.

"What do you need access to?", Georg asks. "Everything!", Karolin replies smiling. She had run into several error messages during the week prior to this meeting and was anxious to get access to the data she needed. "But you do not need everything!" Georg replies. It is not entirely clear what areas she needs access t; the platform is new and Karolin one of the first developers to get access to it. Their conversation is colloquial and relaxed. They have been working together for several years and were transferred to the new data analytics unit together. "Just bring me your laptop. Let's see what we can do", Georg suggests, and Karolin jumps down from the windowsill to fetch her laptop from the second desk in the room. She sits down between the two of us balancing her laptop on her lap. We sit squeezed together behind the desk to see what Georg is doing. She opens her laptop and navigates to the cloud interface. She has to first set up a two-factor authentication which she does with Georg's guidance. Having managed this first step, we all see an error message on her cloud interface. (While authentication was successful, she has no rights allocated to her account yet, what authorizations her account is equipped with has to be determined by the code.)

Simultaneously, Georg searches in the web for the cloud service's permission guidelines on his computer. He opens another browser window that is full of links. He finds the one he was looking for, clicks on it, and the access control code appears. Karolin leans over his shoulder to peak at his screen. Georg opens a policy code window and copies in some code. He leans to Karolin's laptop and her and my head move too. Error message. He leans back to his laptop. "Does there have to be an asterisk?", he is throwing out the question to everyone in the room. He navigates hastily through the code using the arrow keys on his keyboard. "Maybe they want it that way?" he speaks to his screen. He makes a change in the code and clicks "validate" and "save" after the code was validated by the system. A short moment of suspense. A warning overlay pops up asking if he really wants to overwrite. He seems familiar with the overlay and clicks "ok" without leaving time for me to read it. He reloads the tab on Karolin's laptop by pressing the F5 key on her keyboard. A moment of suspense. Again, an error message appears. But this time, he reads it out loud. He moves back to his laptop. "Let's try this". He changes something in the code, clicks "validate policy", clicks "save", clicks "ok" at the warning overlay, and reloads Karolin's tab. After a moment of suspense, again an error message appears. He moves his head back to his computer: "Let's try this". He repeats the procedure several times commenting on his counterpart something like: "He is clever".

Validate, save, ok, reload.

[over and over again].

Validate, save, ok, reload.

He had just pasted a different code snippet into the window when he presses the save button and closes the window. I am confused, and so is he. "Did I just close the window?" "Yes", I reply uncertainly if he was talking to me now. "I am too tired to do this now", he says and laughs it off with a hint of resignation. He begins with the process from the top: He types in a Google search, finds the first cloud interface homepage, clicks on it, and finds the policy code. He continues his routine: edit code, validate, save, ok, reload.

At some point, he stops and wonders: "Where does this explicit deny come from?" Now Karolin has a response: "Possibly it set in the role definition, and it overwrites every policy change you make." "That could be it!" Georg replies with a hint of a surprise since Karolin is no expert on access policies. He begins studying the code using the arrow keys. He leans forward to be closer to the screen, using his left finger to move over the glass screen touching just below each line of code, following every step in detail. His other hand rests on the keyboard, ready to make changes.

I ask what "explicit deny" means. He explains: "It's a security thing. They force you to have two-factor authentication, but you have to be able to see the dialogue in the first place. If there is a read-only policy by default, it overwrites all others, hence an explicit deny. The cloud user by default is not permitted to do anything—somewhere in there is a read-only".

He explains while moving his finger on the glass through the lines of code. He stops suddenly and clicks several times in one place with his mouse. It's the place he had identified with his finger that came to a halt on the glass. "Why can't I change it now?"—"Ah", he answers his own question while moving back up to the beginning of the page, "I am not in the editing mode". He activates: "change policy", searches for the spot in the code, and double clicks. He deletes one line, validates, saves, clicks ok, and reloads on Karolin's laptop. Instead of an error message, the rather boring interface appears. When she sees that Georg had found a functioning permission code, Karolin shouts out: "Yay!" Both begin to smile and chat briefly about what the right code was to achieve this outcome.

The episode demonstrates the grind and tedious material work that it needs to configure access policy *in a secure way*. Georg tries to give Karolin the "right" amount of access that does not produce an error message. Georg makes several entries into the code and employs Karolin's computer to see how the changes affect her interface. It is a lengthy process of going back and forth, reloading, and changing little pieces of the code. Mistakes happen in between: Georg closes the window and reopens it, again finding the code and clicking through the settings.

The episode illustrates the bit-by-bit work to find a policy code that gives Karolin access, but not too much. Security emerges in an ambivalent shape here: It is not preset as to what authorization Karolin should get to be *"not everything"*, but enough. What is secure is being negotiated between Karolin, the laptop, the code, and Georg's experimentation.

Digital security in this episode requires probing and exploring, trying and failing several times. It is crucial to understand that the several attempts and error messages are important for Georg to learn that the code is *"clever"* and to find out about the *"explicit deny"*. It needs several attempts and Karolin's help to find the right position in which the code hangs together to allow access. Think of Georg's actions of moving back and forth, attempting to make the code fit as a sufficient authorization for Karolin.

There is nothing about this process in the documentation for the cloud service, nor in any of the assessments that we looked at in the previous chapter. Georg must be attentive to the changes the code introduces. He is focused on the code, trying to find what effects individual signs could have. This is not text-book security, it requires experimenting and trying. His work surpasses standardized procedure, it goes beyond compliance. By experimentation, I mean that he tries out different code combinations attempting to figure out which one has what effects, doing and re-doing the policy code several times.

Seeing digital security is tuning work, or care: "try[ing] again, try[ing] something a bit different, be attentive'" (Mol et al., 2010, p. 13). Digital security work requires much of Georg who cannot be too tired as he is concerned about making mistakes like closing the browser window. It requires his concentration to touch the code and testing it until it fits.

Georg was concerned with the access policy code many times during my stay. He had to revise the policy code regularly and it was often reported in meetings or over lunch that someone needed new permission. Georg then had to make time for that in his schedule. Georg had to change the policy code frequently to either allow access to new areas or restrict access that was no longer needed. He explained to me that this was particularly the case at the beginning of using the new technology, but he also stated that he expected this work to continue even beyond this beginner phase. Setting policy code did not always take as long as it did for Karolin, but the process was frequently accompanied by Georg testing the effects of code on people's laptops.

While this was necessary to set access code in a fine-grained way (as desired by security regulations), Georg was also committed to learning about the new system, and he was convinced that a fine-grained setting would be best; to the annoyance of his colleagues who had to contact him every time they needed more access. One of Georg's colleagues told me that Georg may take particular pride in setting very fine-grained access policies and being accurate about the securing of the system. His choice to "explicitly allow" very few actions would be praised by the security officers in his firm, but this resulted in the fact that he was often contacted by his colleagues and setting the policy code remained an ongoing effort. To Georg, this meant working overtime to catch up with fine-grained access control policies and his other tasks. He more than once defended his aspiration to make security controls very fine-grained in order to not allow too much.

Georg approached access control rather in a "project-like" manner, getting involved with everyone's authorization individually. However, this was just a solution for the first days of the platform, and once more people would need access, he was planning a more advanced access control matrix—and a second developer to apply a "four-eye principle"—it was important to Georg to remain in close conversation with his colleagues on what is *"really"* needed (not just *"everything"*).

This continuous work was hard to account for in official project planning. It was difficult to anticipate when Georg would have to make a change in the policy code and how long it would take him to figure out the right code. Hence, planning for such tasks was difficult. In Karolin and Georg's team, security was defined as a "non-functional requirement" in the monthly task planning process. As a non-functional requirement, it was not part of explicit "user stories"—specific functions or requirements that organize the team's processes.

User stories are allocated with particular time frames (in workdays). All team members should be capable of working on user stories, meaning that they can shift their resources between tasks when needed. During the monthly sprint planning, the approximate time was estimated that a story should take, which was channeled into a calculation of how many new tasks the team could handle, or what tasks to shift resources from. Each team member would estimate the time for a user story and if there were large discrepancies a middle ground was discussed. This process was rather straightforward, and the team mostly agreed about the time needed for individual stories. Security issues, however, were treated somewhat differently, causing moments of hesitation and awkwardness during the planning meeting. Nobody could estimate well how much time they would need. The team ended up trusting Georg's estimate also leaving him some "free" time to allocate at will, for example when it became necessary to re-configure the access control policy code.

Digital security tasks—other than common user stories—could not be swapped among team members, nor were they listed in detail on the formal progress report. For digital security tasks in the team, rather broad categories were found like "establishing access" or "developing operations", which was identifiable to Georg and calculated in his schedule.

Digital security work is hard to account for in informal time schedules. It cannot be confined to user stories that can be finished (or at least have milestones). As a residual task that was not captured by the progress of particular user stories, it evaded formal time allocation and collaboration methods. It cannot become part of the usual timekeeping and sharing of tasks as it relates to shifting requirements and needs and a particular skill set. Digital security work was temporally unbound—it was not a one-off task that can be measured but required continuous engagement. It was hard to plan when a security issue would come up or when Georg would have to adjust the access policy code.

In the logic of choice, time is linear, provision of facts is followed by a choice which is (neutrally) followed by technical actions. But approaching digital security with care, this linear time "twists and turns" (Mol, 2008, p. 62). There is not one moment that exists where a decision is made on which all actions consequently follow. Instead, problems emerge as you materially engage with another, and when tackled new questions arise. "In the logic of care, it makes no sense to put arrows between events and order them in a linear manner" (ibid.). Caring attunes to the "many vicious variables" in a continuous process (ibid.). While user stories are often contingent on linear work (do one milestone and then another), security work did not fit in this confined logic. Rather it was ongoing, continuous, and circular. Approaching digital security from the logic of care draws attention to such nonlinear processes that are hard to capture.

The overall development procedure also contributes to this ongoing character of digital security work: One of the team members explained that products were currently released as "permanently beta", abiding a final version that is ever complete. This has consequences for digital security that cannot be tested before a product is released but remains constantly in tension with development. Particularly in agile development forms like this employed in the team, development cycles are not finalized by a round of testing but rather by constant and distributed publishing of new versions of software.

Previous narratives of digital security do not account for the trying, failing, and continuous engagement that Georg showed in the team. Putting the literature on care to use here points to the continuous work of attending to digital security mechanisms that cannot be perceived as a one-off and often do not fit in the tight project management agendas. The material work of digital security encompasses tinkering and trying that falls through the grids of formal accountabilities as established through project management techniques. It needs creative strategies (like "free" time, or broad user story titles) to circumvent the formal accountabilities to do such digital security work. This makes it invisible to the Agile Development process but also largely invisible to security officers who might not always see the complex and continuous work that developers like Georg put into caring for their colleague's access policies.

Care is not a task in itself, as rarely is security (Georg has other tasks). Access control issues came up in between and during other tasks, disrupting and intervening in linear times. Care rests upon experimental perseverance rather than solving and moving on. As a matter of care, digital security work is continuous, tinkering, and failing; and it cannot be pressed into temporal boundaries. It requires a shift in perspectives to account for more than linear workflows to account for such work.

The previous section has argued that digital security is a material practice, rather than a mandated formal procedure; and that this tends to be invisible and hard to account for. The next section introduces another dimension of caring that becomes apparent in the same episode as above. Georg's work is morally charged.

Setting "Good" Access Control

In the previous section, digital security was not simply "rolled out" into policy code but articulated through tinkering practices. Digital security was also subverting formal organizational time-allocation and team management. In this section, I want to argue that this "messiness" of digital security work also involves awkward and uneasy decisions, invoking the moral dimension of caring. Digital security comes with a set of moral imperatives (Act compliant! Be secure!) that are a lot more ambiguous and impure in practice. Put differently, digital security is hard because one must engage in practical and moral dilemmas where there is not one but several "goods". Digital security engages in conflicting moralities, and it sometimes produces inconclusive and undesired outcomes.

> "Now, can I try something?", Georg asks, requesting Karolin's permission to use her laptop for a little longer. "Sure, after I document [the process] for Alex", Karolin replies without looking up from her laptop that she had placed carefully on her lap again balancing each tap she makes. She takes screenshots and copies them to a Word document where she takes brief notes on what to do in each step because among her colleagues, she is the first to get access. "We need to play a little and learn how it works", Georg explains to me. He turns to his code window.
>
> After a while, Karolin places the laptop on the table next to Georg with some warning words to not misuse her credentials. She says it smilingly

> and with a friendly tone. She adds: "People will say Mrs (she says her last name) has deleted the lake! I see where you are going". The three of us break into laughter. Georg begins his previous routine, entering code into the window and reloading Karolin's window.
>
> But the situation has changed, Karolin's joke has attested to that. Whereas before Georg was the one limiting Karolin's access and making clear that she "does not need everything"; now Karolin is the one reminding him she is granting him her credentials and is formally responsible for what is being done with her account.
>
> "What if I put an * in here, allowing everything?" he wonders. "The security people would kill me!" he says laughing, reflecting on the strict control of the department's security officer. "But we have to play around and learn", he addresses me apologetically. He keeps tinkering in the permission code for another 15 minutes, occasionally checking Karolin's laptop and thinking aloud: "Oh, wow, now you can do what I can do, that is not good!"

Georg's jokes about "cybersecurity people killing him" resonates that security officers would not want to see permissions being set so vaguely. Georg and Karolin reflect on the morality of digital security: they point to being blamed for "bad" security by "the security people" or "people" who see that Karolin's account is being used.

First, what becomes apparent is the moral judging entangled with digital security. Both a normative good and bad exist to be acted upon. However, during their experimental tinkering, Georg and Karolin also reveal the ambiguity of this morality. They transgress rules and reflect upon that by referring to others' moral judgment. They subvert forms of digital security where responsibility can be delegated and is confined to individual actors.

Attending to this episode from the lens of care points to the alternative moral register that emerges here. This is not about Georg delegating to security officers, or Karolin delegating blame to Georg, but about acting in-between obligations, coming to a new situated morality. The situated obligation is to play around and understand the code (potentially to be better in setting access code "right" in the future). Georg and Karolin's reflections make the conflict of moral orders apparent.

This is particularly visible the moment Karolin passes over her laptop to Georg. First, she has her laptop on her lap/her fingers on the keyboard and can be made accountable for what is written on the laptop, due to it being identifiably signed up to the system using her credentials. She needs to be compliant with the security rules, and if she isn't she has to account for it. Now, at the moment she hands the laptop over to Georg, she explicates a joking warning. Although she trusts him and he had been able to giving her the "right" amount of access so far, now obligations shifts: Suddenly, she warns him not to "delete" anything. She compromises her accountability to "the system", which suggests that she is the one making changes to her system when being logged in. The handing over of the laptop demonstrates how responsibility shifts through the material composition and the code policy attached to it. Before, when she handed over her laptop for him to tinker around to find the right access code, her account was not equipped with any rights. Now that has changed and passing over her laptop also means passing over her permissions. But with passing over the laptop, she does not delegate responsibility over to Georg; she does not warn and judge him. Rather she acknowledges the transgressing of formal obligation, making up a new register between being free of or delegate with responsibility for digital security.

The two show that there are modes of securing that go beyond who is formally responsible. Georg reflects upon this fuzziness when joking that security people would not like to see him doing that. Georg and Karolin do not issue a moral judgment; through their joking, they resonate with the existence of formal obligation but putting their situated effort and relations at the foreground.

Georg and Karolin experiment with access rights and reveal, jokingly, the moral ambiguities of digital security. The laptop enacts the shift between clearly appointed responsibility (Karolin and her laptop) to complex forms of becoming obligated and responsible (Karolin giving her laptop to Georg). Leaving it with Georg to continue his tinkering gives Georg the access rights that Karolin has. And thus, she reflects on her moral obligation as a user of a system and the fact that she would be liable if Georg did something wrong. But she does not do so by setting a normative rule: never use my laptop, or by judging Georg, but by engaging with him and trusting in his competencies. From the viewpoint of security officers, this needs to be condemned as it does not follow a strict accountability—who is responsible here? But, seeing that security work—as it is done here—entails more than delegating responsibility, but trying something out together, unfolding complex obligations to the system and each other as colleagues. Georg and Karolin's joking shows they tolerate and deal with ambiguity rather than find a permanent solution or "moral closures". Attending to what is good in this episode cannot be done without connecting to the specific situation and material setting. What is good creates a realization of the setting rather than that of a general norm.

Looking at digital security work from the lens of care makes the conflicting moral registers between formal delegation and responsibility and local, collaborative and touching obligation visible. Shifting moralities visible. During the material work of setting access, we can understand better how moral judgment and local forms of obligation to each other, and the security of the system emerge and intersect.

Georg and Karolin engage with normative morality and open conflict with it when developing a "situated ethicality" that responds to the need to experiment and tinker around (Puig de La Bellacasa, 2017, p. 115). They do not leave formal obligations aside entirely but remain responsive and affected by them. Care can account for such ambiguity and unruliness.

Georg's piecemeal work to set access control not all at once but as continuous, fine-grained, and ongoingly, shows his attachment to the technology and his obligation to the risks that come with access control policies. Georg and Karolin decide what is "good" access not by strictly complying to the rules, nor by freeing themselves of them entirely. They do so by applying care. Care refuses to conceive of good as general means. It is committed not to judge but to apply a circumstantial and often improvised appropriateness or fittingness. Its "non-normative ethics" (Puig de La Bellacasa, 2012, p. 211) acknowledge moral ambiguity and open-endedness: "Good" and "bad" are never settled in the logic of care, they are exercised in practice. Caring is ongoing and requires a long-term commitment. It does not end when Karolin could be blamed or Georg is punished by the security people, it deals with the delicate morality of the practice by developing situated obligations. But it also—and this is what

Georg and Karolin do—acknowledges they might fail, and things might go badly. Care is not naïve, it actively and consciously acknowledges the consequences of its moral work but continues anyway as it has to deal with material constraints.

Caring for digital security is "not making value judgments, but engaging in practical activities" (Mol, 2008, p. 86). As Mol argues, care is not about locating the mistakes of the past: "The logic of care does not impose guilt, but calls for tenacity" (Mol, 2008, p. 91). Care does not take the stance of blaming developers for "bad" security, blaming Georg for deleting the lake or Karolin because she wants access to everything, but gets involved in the moral ambiguities of their work and wonders what is best for the situation at hand. Or, put differently, the morality of care is not about blame and shame (Puig de La Bellacasa, 2017, p. 95), but about morale and dedication. Since doing digital security must contend with the fact that security is never absolute or final, like care digital security practice, it "seeks to lighten what is heavy, and even if it fails it keeps on trying" (Mol et al., 2010, p. 13).

To care is moral, but not normative in the sense of condemning "bad" security practices. Caring is not about making judgments and holding people accountable against normative standards. It is also not about finding a binding agreement or definite closure. Rather, caring keeps things open-ended. To care is engaging in efforts to do good without having a final answer. Caring acknowledges moral heterogeneity and dares to remain undecided. Georg's work of going back and forth, trying and checking, while reflecting jokingly about moral and organizational standards on which he trespasses show such efforts to stay open. The jokes are almost apologetic, they address the outsider ethnographer that may or may not be here to report back to the digital security officer how well the team complies with digital security regulations. But Georg and Karolin dare to transgress that. Caring includes such daring, such as keeping open, even if it could be punishable under normative standards.

Georg's way of establishing security is far from obvious. He does not stick slavishly to company policy but breaches it temporarily, on his search for a secure solution in this particular case. As mentioned earlier, such solutions were improvised and temporary. Georg is a skillful and experienced developer who is trusted by the team and is knowledgeable (and sometimes annoying) in his security recommendations. While this is what security officers might want in a local security gatekeeper, his work is not in complete conformity with security guidelines.

I want to add a little more about Georg's position in the team. Georg was persistent about digital security in the group. Other team members admitted that they thought Georg might overdo it sometimes with his constant worrying; for example, Georg regularly noted in meetings that applications were security-relevant and reminded his colleagues that security should be respected. He critically commented on some of his colleagues' ideas because they would not comply with security regulations. But he did not do so in a warning or blaming manner but by posing options and possibilities of how to include security in their ideas. Georg's practice is that of a local security expert, he did the translation work necessary to put security regulations into practice.

Georg was a particularly skillful developer as he was constantly worrying about digital security in the team. His role was usually that of security advocate in the team, without being formally responsible for it. He was respected by his colleagues, and they had very good relations. Some of them had made a request to the management to work together in the team. Georg reminded his colleagues of security functions and stood by when they tried out new technologies that were implemented with security mechanisms they did not know. Before the episode I cited above, Karolin had trouble using the two-factor authentication and Georg helped her patiently and without hesitation. Georg was the "security guy", but he rarely blamed and shamed his colleagues for bad security decisions. Rather he was himself "wrestling the three-headed hound from hell" that he called IT security, information security, and data protection. Georg was highly respected in his team and one colleague jokingly told me that he thinks Georg has a "secret life as a hacker". Georg's obligation emerged from his troubles of dealing with mandated digital security regulations. He took pride in doing fine-grained access control policy and his colleagues suspect him of having a secret hacker life, attesting to his personal commitment to the matter. Georg's security advice was always received gladly, and he never acted in a judging way (if they did not know how to activate two-factor authentication, or when an intern lost her credentials to the Data Lake). He gladly issued new ones, often acting a bit annoyed and busy but always keeping his office door open for inquiries.

Understanding Georg's and Karolin's obligations to security in this example and Georg's practices beyond the episode need an understanding of the specific practices and settings in which digital security is attended to. Looking at Georg and Karolin's work as a form of care allows us to investigate the intertwined and shifting responsibilities that cross formal organizational boundaries that are part of digital security.

Georg's concern for security was neither completely compliant with security guidelines nor entirely non-compliant. Georg worried about digital security, and he struggled with it and referred to it often. This worry became evident to me when he explained that he had developed an acute reminder in his work:

> We sit down on Georg's computer as he wants to show me something. Mysteriously, he opens his Microsoft Explorer browser window and then his Firefox browser window. He had installed a red-coloured theme from the movie "Deadpool" to the Firefox browser which turned the window into a cherry red colour. He asks me: "Why do you think I did that?". I look at him and laugh: "Do you like Deadpool?". He smiles at me wisely and replies with a big smile. He begins to explain that Powino had a security policy that forbids opening websites with an HTTPS://certificate. Knowing that this is a more secure certificate than the standard HTTP://, I appear confused, so Georg continues: "They assume that it is more secure to prohibit actions in the browser than to rely on websites to use HTTPS properly". "But", he continues with a clever smirk, "this prohibition only applies to Microsoft Explorer", he smiles widely now. "That is why I use Firefox sometimes if I have to open HTTPS sites. And the red colour helps me see immediately what browser I am in".

Georg employs a visual trick to know when he complies with and when he subverts company standards. While also pointing out that he uses the other browser to open HTTPS://URLs, which would be considered as high awareness for security issues. Georg worries about digital security, not in the sense of rule-abiding, but by experiencing insecurity more explicitly. He set the browser to a red color to remind him when he's acting outside company policy.

Remember that he searched through the code in the example above by running his fingers on the screen to follow each line of code with his eyes. His care work is done with his entire body. Trying to stay attached and attentive to security. His worrying comes in the form of touch: being attentive to and able to respond to something.

Obligation through care does not come through abiding by abstract ethics but by "touching" and engaging with the dilemmas and material intricacies of concrete situations (Puig de La Bellacasa, 2017).

Georg and Karolin show that their obligation to digital security is different from a regulator or security officer's normative stance. Their obligation departs not from the standard, but from local practice and close relations. Georg worries about applications that colleagues develop, advising them on ways to include security. But he does not do that in the from a remote, distant, and blaming position. Georg and Karolin experience the morality of digital security directly as they are blamed for deleting the cloud or that they could be "killed".

Puig de la Bellacasa notes that thinking about care as a way of "touch" emphasizes vulnerability in relational ontologies. None are unaffected by the other. But the vulnerability here is nothing to dispose of or think of critically but of necessary intrusiveness and appropriation. "[W]here boundaries of self and other tend to blur", caring relations are built (Puig de La Bellacasa, 2017, p. 119). We can translate this to digital security. Where digital security is configurated, the boundaries of humans and technologies and between inside and outside blur. Being in relations, intra-acting in webs, being vulnerable is not to be disposed of as they are the essence of being in the world. Combining digital security—largely grounded in assumptions about difference, boundaries, and seams—is challenging. What care helps us with is to understand how relations and being related are nevertheless present in digital security work, and neither can they disposed of. Fragility, vulnerability, and relationality are essential to technologies as much as to human beings. Care refers to the fragility of objects that require attentiveness, not imposing ideal goods and control, but caringly playing with what is good for digital security (Heuts & Mol, 2013, p. 138–139).

What constitutes care for digital security in these examples is the "careful 'reciprocity'" between Georg, the code, the interface, Karolin, their colleagues, and the security officers (Puig de La Bellacasa, 2017, p. 120). Puig de la Bellacasa argues that this sensorial materiality, the sense of touching, shows a quite different obligation to matters than distant norms and rules. She identifies a "quality of caring obligations" (ibid.) that differ from political obligations that can be encountered with sanctions and punishment. What characterizes such caring obligation are rather "multilateral circulations of agencies of care"—one cares for one another, and obligations shift within situations. Georg and Karolin's exchange is exemplary for such multilateral circulations. By handing over the laptop for Georg's explorations, Karolin cares for her colleague, learning about what effects code can have. They subvert the security officer's guidelines to enable this mode of experimentation and tinkering. Only by getting in touch with and becoming accountable to each other, care obligations evolve.

Care circulates multilaterally and collectively. It is shared and shifts constantly: Georg caring for Karolin's access, Karolin caring for Georg's will to learn about the code's responses. But care does not expect to be symmetrically reciprocal. It gives without expecting a return. Georg's will to "overdo" security mechanism and his constant troubling and worrying about security in the team is not responded to by reciprocal care. But since he does not make it out to be the other's choice to listen to him or not, care disrupts the moral compound of obligations and responsibilities that have prevailed the agency of previous morality (Puig de La Bellacasa, 2017, p. 121). His constant warnings and worries don't come in the form of expert stories, but as constant and situated engagements.

Care in Feminist Studies and STS

Theories and considerations about care are various. Although deriving from feminist sociological explorations of mother-child relations (Gilligan, 1982) and healthcare settings (Mol, 2008; Latimer, 2000), care has also been discussed in relation to ethics for animal rights and farming practices (Law & Lin, 2010; Singleton, 2010) and, closer to the inquiry in this thesis, for analysis of politics in technologically saturated worlds and ethics of design and computation (Martin et al., 2015; Houston et al., 2016; Jack & Jackson, 2016; Toombs et al., 2015). All such engagements (and there are many more not listed here) make specific contributions to understanding situated care practices. Taken together they do not form a coherent picture of "good" care but rather point to empirical specificities of what care can be in each example. The investigations speak of utterly different practices when speaking of care, some of which may be even contradictory (Puig de La Bellacasa, 2017, p. 3).

Hence, rather than providing a definite idea about what care is, I want to begin this chapter by pointing to some of the questions that care researchers have asked in their fields. Questions that are relevant to the study of digital security.

For Berenice Fisher and Joan C. Tronto (Fisher & Tronto, 1990), care is composed of two related dimensions; maintenance or what is commonly termed "care work": The material doing that is supported by objects and infrastructures, and is contained by physical possibility (e.g., psychological and physical abilities, etc.), and a pursuit of good, an effective drive toward doing "better" or making something "as good as possible". This emphasizes that care takes a moral stance, not by distant judgment, but one that involves hands-on agency, getting engaged in, and being constrained by material forces. How does normative judgment change when being confronted with concrete material constraints?

I want to present two examples to make this point clear: A parent cares for their child and wants them to live as well as possible, they do so to the best of their (physical, financial, emotional) abilities. Eventually, part of that care is to let the child stand on its own feet and be independent. This is a classic description of care that emphasizes that caring can entail stepping back and keeping the "right distance" (Puig de La Bellacasa, 2017, p. 5). It also makes clear that caring is not all good and easy, it is hard to care/let go and do what is best for a child. A clear moral undertone accompanies the story where giving one's best and doing good comes as a distinct norm that is hard to live up to. Caring well for a child can mean very different things in different contexts.

Another dimension of the intertwined material work and moral obligation becomes transparent in the context of healthcare: A healthcare worker tries to give their best to care for all their patients but must spent their time wisely with one or another. They know that things for some patients will not end well—will they attend less or more to them? Their work schedule puts pressure on them, and so possibly does the (mediocre) payment they receive for their work. Their care work is paid, yet how they attend to patients—take extra time here and there or prepare a favorite tea or meal—is another matter.

It is important to understand the situated relations of moral obligation and material work in the two contexts. In one, it is kinship and love that attach the caregiver, while in the other, it is professional pride, humanity, affection, or empathy. The first is unpaid, the other is paid. It is crucial to see how material work and affection and morality are linked together in specific care settings. As Maria Puig de la Bellacasa notes, a worker might want to limit their care and affection to avoid being exploited as waged labor (Puig de La Bellacasa, 2017, p. 5), while a parent might want to distance themselves from a child to foster independence. These two classical examples of care work put an emphasis on the ambivalent quality of care, on moral obligation and affection. What care is means different things in every context, and maybe for every child/patient? But the emphasis when looking at the care practices of both is how material work and moral obligation are linked and how doing good is a situated endeavor.

Mol (2008) describes the practices of doctors who, in partial defiance of medical protocol and standards, negotiate "good" treatments with patients. In her example, patients show problems in fitting glucose measuring into their daily life. Instead of blaming diabetic patients for forgetting to measure their glucose levels or practicing "bad" eating habits, doctors and nurses engage in the work of finding the "right" measures for these patients. A patient that might want to visit his daughter's wedding may realistically eat a piece of wedding case, although he is not supposed to. The doctor may then suggest an extra dose of insulin. While this is not an ideal solution, it is a good one for this patient. Doctors and nurses engage in communication with their patients to find out what concrete circumstances keep them from complying with medical advice. They might introduce alternative rhythms of measuring or allow patients to neglect their diet once in a while, not because it is medically advisable, but because it suits their practices and is realistically easy to achieve. In short, rather than simply follow medical procedures, doctors and nurses care, and even compromise formal advice.

Mol contrasts caring with giving a "choice". Offering a choice means laying out options to the other without acknowledging difficulties of choosing or practical consequences of one choice or another. The logic of choice conveys the idea of rational and informed decision-making where the one giving the choice stays unaffected by the one making the choice. The logic of care operates differently. Rather than removing oneself from the decision, doctors and nurses get involved in making imperfect recommendations, to improve their patients' lives (Mol, 2006). In the logic of care, failure is possible and can affect both.

I am putting forwards these examples to suggest that some of these descriptions can be mobilized for digital security. How to define what are good security measures in a given situation? What if reluctance to comply is not answered by blaming but by finding compromise and trying differently? And how can caring for digital security produce specific and locally relevant obligation, attachment and responsibility?

From the first considerations of care in feminist studies to Mol's empirical examples of doctors and nurses, many caring acts remain invisible and cannot be captured by formal responsibilities and liabilities. All the examples above attest to that; parental care work is less visible than other forms of work (i.e., labor); and those caring activities, such as staying longer at a patient's bed or finding alternative treatments, are hard to account for through formal schedules and treatment plans. Care work is invisible. Questions in feminist studies, consequently, included how to shed light on those practices that lack institutional visibility, and how to how to organize social life differently in order to better account for and value care work.

But caring cannot simply be rendered more visible in order to "implement" more care, or to better account for it. The invisibility of care must itself be treated with care. Not all invisibility is bad. Caring includes the possibility of subverting official norms and guidelines. Caring can be unruly. Taking the time to speak to a patient for longer can be in conflict with official norms that calculate the same time for every patient disregarding their mental health and wellbeing.

Lately, a more generic discussion has taken place that detached care from its origins in feminist and healthcare context toward a "'generic' doing of ontological significance" (Puig de La Bellacasa, 2017, p. 3). For Fisher and Tronto (1990), care has to be understood as a "species activity"—or as everything we do to "maintain, continue and repair" the world we live in (p. 105). This world includes ourselves, as much as social relations, environments, other species, and all that sustains our life. Care lies in practice, it is something that we "do", but caring is also political, an ethical commitment to do good (Puig de La Bellacasa, 2017, p. 5). The doing is infused with questions of caring for whom and when, and how caring can be accounted for, become hurtful, etc. To care means to be attentive and to contemplate people and things tirelessly and patiently. Care work is a "persistent tinkering in a world full of ambivalence and shifting tensions" (Puig de La Bellacasa, 2017, p. 7). Care work responds to ambiguity, to imperfection. It does not leave patients alone, but negotiates with them, makes compromises, and bends the rules. It "seeks to lighten what is heavy, and even if it fails it keeps on trying" (Mol et al., 2010, p. 13). Care takes "decay" and possible ends into account (Cohn, 2016). It assumes "vulnerability as the 'natural' state of things and not as a temporary deviation from the healthy normality" (Denis & Pontille, 2013, p. 14).

Care practices cannot be standardized; they derive from watchfulness and attention in acute situations (Mol et al., 2010). Care draws on a certain engagement with objects, "a bodily commitment" (Denis & Pontille, 2013, p. 14) or "touch" (Puig de La Bellacasa, 2017, p. 95) instead of an abstract judgment.

From such considerations emerges a specific "ethics of care" (Tronto, 1998) that evades objective evaluation and distant judgment prevalent in mainstream ethics. Ethics of care can account for pragmatic and situated moral ambiguity; specific, personal, detailed, and responsive to small actions and reactions. Care ethicists attempt to understand and get in touch with the local practices of groups rather than judge their actions from an abstract normative stance. The ethic of care seeks to "illuminate the relationships of small components, one to another, within great systems" (Nowviskie, 2015). Such ethics sees caregiving as a duty of deep empathy to the specific, while maintaining—or being capable to respond to and perceive—systemic effects (Nowviskie, 2015).

That suggests that caring entails a specific "modality of handling questions to do with good" (Mol et al., 2010). In the logic of care, ethics rest in practice, not in regulation or abstract judgment. Obligations emerge from doing, not from distant appointment. However, blame and shame are not foreign to practices of care, rather they co-exist, leading to unconformable and unsettling 'disconcertments' (Verran, 2001). Helen Verran uses this notion to emphasize the discomfort that she experienced seeing her own and other epistemic systems meet. She argues it is a sensation worth exploring for ethnographers. Translating it to digital security in this case entails paying attention to the moral ambiguity involved in acting upon security, engaging with diverging goods in practice. Doing good comes in as many registers as systems that meet. Rather than overcoming the uneasiness of digital security, it is worth "staying with the trouble" by unfolding the struggle to care for digital security in practice (Haraway, 2016).

Handling questions of good in the logic of care is a matter of physical attention, attention to the local and specific dilemma. It can mean a wide range of practices and can include doing nothing. Caring can mean to leave things be, to not touch upon them (right now). It does not stand in conflict with neglect. Caring leaves spaces of tension, hesitation, and failure. By doing so, it allows subverting asymmetrical power relationships without the expectation to convey "better" morality.

Care practices can subvert or counteract other registers of work, e.g., standardization or regulation. In such moments, care becomes conflicting and contentious. Caring thus also entails subtle forms of resistance and working in out-of-view ways. This doesn't mean that care is free of power, nor is care always feel-good and naïve. Care is susceptible to conveying control. Caring can be paternalistic; one exercises power by determining what is "good" for someone or something else.

While shifting from insecure to secure is often considered a simple decision—of applying the "right" code, using a "good" password, etc.—this chapter argues it is rather a matter of care, in which alternative morals emerge: "In the logic of choice, a good decision depends on properly balancing the advantages and disadvantages of various courses of action"

(Mol, 2008, p. 61). In the logic of care, balancing is a material act of attuning to such vicious variables of everyday life. Speaking of compromise and negotiation then becomes less an act of two prior specified positions—a bit like it was in the last chapter—but more of material attuning to, and seeing where and when, problems appear. The two (or more) sites are not predefined, e.g., developers on one site and security officers on another. Attuning to variables that shift and turn in the process also means that caring does not completely evade, nor entirely comply with what is considered "good", "healthy" or "secure", rather it describes what happens in the meantime; in between being "good" or "bad". In such ways, care challenges other ethics based on justice, economic exchange, or choice (Gilligan, 1982; Mol, 2008).

Caring is not done in isolation and is not only a human activity. Machines, objects, and devices are also part of caring work. But being concerned with others' well-being goes beyond instrumentalist relations commonly held to perceive our interactions with objects. Care does not simply "deploy" objects, it speaks to an attentive "being with"—to which the image of caring for a patient or child attests. Care is carried out with help of objects and technologies, but they are also cared for, and care for us. As Clémence Pinel et al. (2020) argue, scientists care for their research data; or Marisa Cohn (2016) illustrating how engineers care for high-tech devices as they (both) get older. The lives of others—including objects—are secured and sustained through attachment and touch and over time.

To sum up, care work does not produce general solutions, but local spaces of possibility. Caring can respond to uncertainty and ambiguity, not with a concluding strategy but by putting itself into proximity with the issue. Caring can mean subverting formal norms and finding local alternatives; in that way, it responds to ambiguity and carries a moral other than rule-abiding. Caring means remaining attached to a matter, worrying about it, and being affected. Care comes in the form of "practical tinkering, of attentive experimentation", of trying one's best, being able to notice and respond to each other (Mol et al., 2010, p. 13). Caring suggests enduring work that seeks to improve but does not always succeed.

The chapter asks what caring for digital security looks like and how questions of moral obligation, ambiguity, imperfection, and accountability are addressed in the material work of digital security. What if digital security is approached as a matter of care rather than choice? What if we look at the tedious and often imperfect work of attending to digital security rather than pose moral imperatives? How to facilitate caring for digital security without falling into normative judgment? Caring for digital security will offer yet another embodiment of care, pointing to yet again different practices and tensions that emerge while caring.

Implications 3 and 4

Setting access control policies is a continuous process that requires "calibrated care" in which possible outcomes of a certain code setting are tested and experienced. There is no ultimate good before the situation described here, and the code remains clumsy and provisionary afterward. The tinkering is not naïve: Georg reflects how his experimenting work may be condemned by security officers and what consequences this may have for him and Karina. In the logic of care, this episode can be understood as engaging in conflicting moral registers while having a pragmatic task at hand. In the logic of choice, this episode is aimed at finding a solution, maybe a trade-off, but it would aim at a clear-cut answer to the question of how to allocate access correctly.

In the practice of tinkering, fragility is a source of worries that induce a "specific form of engagement with objects", an "acute attention" without alarmism. Tinkering overwhelms standardized procedures—it dwells in the realm of subversion and work-around.

Perceiving digital security as a matter of care requires close attention to the hands-on work and bodily experience of digital security, on what falls through formal schedules and remains gradual, scattered, ambivalent, and contested. As care, digital security constitutes as a constant worry. This worry is disputed, conflicted, morally charged; but it is not distant and judging. Digital security, in this perspective, remains fragile but the lens of care can point to what happens in fragile spaces: moral imperatives are answered with subverting and touching forms of affection and attention. Looking at digital security from the perspective of care moves attention to the material work and conflicting moralities. Instead of posing moral imperatives, it produces local obligations that emerge from proximity and touch.

This enacts digital security, not as a normative rule but as a practical process contingent on specific situations and entangled with alternative moral registers. I want to foreground in this final section what is specific about caring for digital security and what questions the lens of care offers in the study of digital security. I will attend to the latter first.

Implication 3: Digital Security as a Matter of Care

The examples in this chapter have shown digital security work as transgressing and going beyond formal tasks and mandated obligations. Digital security has involved constant worrying, visual tricks, reminding, tweaking project plans, playing around with code and trying out, temporarily stepping over organizational rules. Such care work is notoriously invisible as it is in-between and beyond formal accounts and has to partially stay that way. Putting Georg's work into a formal framework, e.g., allocating specific times for tinkering work in the project management could lead to him missing out on other tasks, or tinkering with access control policies when no one needs them to change at that particular time. Care work eludes such rigor. Perceiving digital security as a matter of care entails emphasizing the practices that are needed to "meticulously explore, test, touch, adapt, adjust, pay attention to details and change them until a suitable arrangement [material, emotional, relational] is achieved" (Mol et al., 2010, p. 16). I have discussed the dimension of meticulous work and the resulting ideas of a "suitable arrangement" that captures both material configurations and people's relations and collaborations. While this shifts away from making normative stances about how to improve digital security, this does not mean that such care work cannot be supported. Specifically, cognizant of the in-between character of digital security work, one approach could be to avoid delegating responsibility for digital security to individual actors or specific groups and refraining from blaming and shaming them. As some fluidity or fragility is necessary to do digital security care work, it can be supported by perceiving local practices and adaptations, rather than undermining them. Rather than enacting taboos, digital security research should find modes of fostering experimentation and tinkering, allowing bonds and attachments to be forged among developers and digital security.

Caring practices appear rather vague and elusive. It is not easy to facilitate them from above or from a distance. Top-down enforcement "to care" undermines the local and embodied forms in which it emerges and is sustained. Although it is hard to explicate what care means outside situated practices, it is possible to foster caring relations. Siegfried and Simon's examples above have shown that from the position of digital security officers. While in practice compromise, negotiation, and bending the rules is common, digital security research has rarely captured them. Perceiving of digital security as care pays attention to the informal accountability and fragile and preliminary ways of achieving it; it evades blaming and shaming and encourages tracing situated moral obligations.

It seems straightforward to say, "We shall secure with care", or "Developers should care for security!". Care in this way is a reminder to do security "properly", something that the notion of care as introduced here refrains from. This statement could come from one of the security officers Georg and Karolin are afraid of being blamed by. This security officer has a moral high ground and Georg and Karolin know they don't follow the standard definition of the security officers. But ethics of care, according to Puig de la Bellacasa, are not about making judgments and holding people accountable against normative standards (Puig de la Bellacasa, 2011). It is also not so much about negotiating binding agreements or other forms of "moral closure". Instead, to care is about acknowledging moral heterogeneity and the open-ended, situated character of efforts to do good (Puig de la Bellacasa, 2011, 2017). Judging from afar does not fit with practices of care; they are not adhering to abstract evaluation but respond to the complex realities and ambiguous troubles of patients, users, or developers (Mol, 2008, p. 60). Caring means getting being related to the difficulties of living up to standards.

Mobilizing Mol's and Puig de la Bellacasa's work on care, what the imagined security officer does—choosing the moral high ground, judging from afar without getting in touch with the code or the daily life in the teams—is not care, but follows the logic of choice: You should do security with care or you are not doing it "right". Caring is not proselytizing. For what it attends to does not work in the register of right or wrong nor good or bad, but as a process, elusive, fragile, and improvised. Care does not blame or judge, it sits there and tries its best with what it has at hand. Georg and Karolin care—not by finding the "right" solution but by defining a locally improvised good and staying committed to its ambiguities.

While user behavior in existing digital security research is usually found to be erratic or careless, Georg and Karolin bring more nuance to it. The way in which the two approach digital security is not erratic but skillful, with reflection on the fuzzy morality their actions entail. Nevertheless, their practices conflict with what digital security officers would like to see. From the empirical episode, we learn that caring forms of digital security require constant attention and moral commitments, they are harder to capture, account for, and support. Digital security as constant and morally ambivalent caring is less easy to confine and measure. It needs methods to better capture such practices, and to think and discuss the way digital security is morally conflicting and continuous work.

By conceiving of digital security in the mode of care, registers of acting upon it are no longer along the lines of defending and attacking but along the lines of how invisible practices configure digital security, obligation, and alternative moral registers worth exploring. Georg's work was not enabled by blaming and judging but by relating to colleagues' tasks, building good relations, getting in touch and being in proximity with specific issues, and finding ambiguous, imperfect, and provisional solutions. He showed a form of worrying that was performed as a constant reminder about digital security and being affected and responsive to digital security matters.

This chapter bridges the gap between a prevalent style of digital security moralizing that resides on blaming and shaming deficit actions, with an understanding of the ambivalent and complex work of caring for digital security. This chapter contributes to the arguments of the previous one: Telling the story of digital security through standards and tests misses the mess, morals, and worries that go with it. Georg worries when inquiring further into the code: Did he make a mistake? Can someone get as much access as he has? Will Karolin's access suffice? These worries render the technology fragile even when the access code is "fixed".

What is good and what is bad in digital security? Experts will have a clear idea about that. Sharing passwords is bad, keeping data for oneself is good. But these clear-cut ethics become more complex in practice, where doing good requires breaching official rules, getting in touch, and developing new obligations. However, expert norms are not left behind entirely. Caring points to the fuzziness of moral heterogeneity.

Emphasizing these practices, I have offered the first step on the specificness of the way in which they deal with questions of good and bad and making them visible to a community largely invested in solving problems. In my examples, I have refrained from blaming and shaming for the purpose of analyzing the modes of action that evade such judgment. This does not mean they don't include judgment or obligation and responsibility altogether. Instead, I have shown how those emerge from concrete encounters, from touch, and by forging good relations.

Putting this in the form of an implication requires digital security research to cultivate the forging of relations and modes of negotiation and collaboration. Rather than blaming and shaming developers, it needs arenas of care for the collaboration between security officers and developers. This is not to say that security officers are the "bad" guys in this example. As demonstrated in the last chapter, security officers do a lot of care work in practice, living with creating alternative forms of security, dissenting from regulatory or legal forms.

It needs practices to improve such practices that temporarily step beyond formal obligations and accountabilities. However, it is exactly this uneasiness that is fundamental to fragile forms: it also needs formal accountability and rigor. I have argued that care can stay and remain open in-between forms of standard normativity and local goods, shifting between and beyond them.

Furthermore, contemporary research on digital security lacks an emphasis on the ambivalent and morally charged practical work of digital security. Where medicine has ethical review boards or committees, the ambiguous moral choices in digital security are neglected and condemned. I have suggested a discussion of the multiple "goods" of digital security and thus an ethics of digital security that resides in care. Care is not indifferent to standardized and normative forms of doing the work, but it gets involved with the hard-to-make decisions, ongoing commitment, and emerging obligations.

This includes considering the power relations manifested in care. While those who care to complain of a lack of power as their practices are invisible and undervalued, caring also conveys power. Care can take the form of paternalism, even a usurping, when it means taking charge or putting normative expertise before local practices and collaboration. In the case of the team here, it is important to note that the new data platform being configured was of utmost importance to the company and was supported by large investment and company-wide recognition. The security departments, on the other hand, were notoriously understaffed and overwhelmed with tasks. Hence, even though Georg's practices show material tinkering and developing situated moral obligations, they must also be seen in a privileged context that could afford some leeway in relation to digital security. It is exactly such relations that the lens of care can uncover and put into context. Blaming and shaming developers is a form of delegation that could be the result of a lack of resources for digital security departments. As others have shown, however, digital security investment should be focused on the right form of intervention, for example cultivating collaboration and care in training rather than taboos and restrictions.

Furthermore, what has not been the focus of attention here, but should be elsewhere, is the role of gender in digital security as well as power dynamics between experts and users. Focusing on professional corporate developers, how users care, and what modes of caring are particularly present in vulnerable communities has been left untouched. This includes questions like: For whom is something in/secure? Who does the nitty-gritty care work of digital security in smaller organizations or among activists? More investigations are needed into those practices of caring for digital security tailored toward state surveillance or reidentification. Also, and even more invisible, little is known about how hackers care. What drives hacking and what material acts and moral obligations are part of (ethical) hacking. With colleagues, I have suggested a few answers to the latter (Kocksch et al., 2018).

Implication 4: What Caring for Digital Security Contributes to the Care Lens

Approaching digital security from the lens of care allows us to draw some conclusions of what specifies care in digital security. I have put an emphasis on the ways in which obligations emerge through care, how forging relations and living through insecurity is crucial to digital security work and that alternative moral registers emerge when looking at digital security as a practice of care. While other dimensions of care could have been in the center, I chose (or, rather, I cared for) those.

Care is "displaced" here to the field of technologies where traditionally military and masculine metaphors prevail (Puig de La Bellacasa, 2017, p. 36). This draws attention less to matters of unequal distribution of invisible work than to understanding modes of attending to technology, and their normative obligations. Digital security practices are often deemed deficit at best, malicious at worst. It is worth attending more to practices of tinkering and engagement to perceive the intricate dilemmas and insolvable troubles of doing digital security work in contemporary technological systems. Accounting for caring practices in digital security makes a distinct intervention in a logic-driven discipline by highlighting attachment and relations rather than instrumentalism or choice logics. Caring for digital security proposes attending to material devices and ethical and social obligations at the same time. It requires the material work of tinkering with machines and worrying about their potentials and outcomes. It demonstrates a mode of being affected with infrastructural capacities that are respectful to contingencies and complexity. While such an approach does not make a traditional feminist argument by decreasing inequality, it points to alternative ways of making technological relations. Such a displacement further sharpens the concept of care. Caring for digital security demonstrates that care can be found anywhere where practices can be found that subvert, uneasily, abstract rules—even in industrial companies. By pointing to digital security as a matter of care, I widen the notion of care thus to include a variety of practices aimed at handling dilemma and "staying with the trouble" when multiple "goods" collide (Haraway, 2016).

I have pointed to the writing of access policy code as a practice that combines material and touching practice with moral commitment and effects. I did that to account for how developers like Georg and Karolin cope with the fragility of technologies, not by overcoming it or blaming others but by getting involved with specific ambivalence and shifting "goods". I have employed the notion of care to emphasize this neglected side of digital security. Care for fragility does not follow a normative good but is prepared to handle the existence of imperfection and insecurity. I have put forward an understanding of digital security as persistent tinkering, continued affection and trying, in situations that are full of ambivalence and intricacies.

What is secure affords carelessness. No, quite literally: The Latin stem "securus" means "without care" or "without apprehension" (Oxford Dictionary of English). But the process of securing requires a great deal of care. Drawing on the feminist notion of care and particularly the work of Mol (2003, 2008; Mol et al., 2010) and Puig de la Bellacasa (2011, 2017), I have argued that digital security involves care: tinkering and attentive experimentation and ethics emerging from getting in touch with complex and inconclusive worlds. Caring for digital security applies tricks to remain at least partially responsive to standard normative rules. As the last implication has focused on the unruly sides of digital security work, which require some alterations in how digital security is researched and trained in corporations, this intervention focuses on implications for the notion of care. Specifically, how digital security fosters an understanding of caring activities as in-between formal and local accountability, and the stakes that are involved in its elusiveness.

By employing care as a mode to describe digital security practices, I contribute to questioning what care is. To care for code, for each other, for global computer networks, for company assets. Caring is moral, yet not naïve and feel-good. Care does not bracket fragility, but "faces up to" it (Mol et al., 2010, p. 10). It can attach itself and be capable of handling the risk. There is care for business, care for money. Caring for digital security in the context of Powino can be care for critical infrastructures and the common good, but it is also care for company reputation, and for monetary profits. This displaced care to corporate fields. Georg's care is humble and honest; he does get in touch and in reciprocity with colleagues and the code, but, nonetheless, this is the corporate field. This points at the usual and uneasy fields of care work. Where are STS scholars troubled in locating care? Certainly, in corporations (even more so in one that is closely related to coal mining and industrial pollution). This also belongs to the notion of care: our uneasiness of displacing it to fields we find notoriously careless. But what needs support are care practices here too. Practices like Georg's; seeking good through experimental tinkering and dealing with ambiguity.

Specifically, this opens the question of how to care for high-tech devices and how caring and innovation are not opposites but tightly entangled. I find we can learn from digital security in this way as it enacts computer systems not by innovating or progressing but by living with what is there—including anxiety of being hacked. The system in question here is brand-new, and it has high hopes attached to it, yet it depends on care work. To digital security, maintenance and no change are on the horizon of action: Georg and Karolin cannot overcome the potentially dramatic effects of their work. Setting access control in this example does not only mean getting the right code but incorporates a way of coping with potential attacks and with pressure from security officers. It requires what has been termed "broken-world-thinking" by Steve Jackson (2014). It knows how to live with fragility and potential breakdown rather than overcome it. I will spend more time unfolding this argument in Chap. 6 when I attempt some conclusions on fragile computing, but keep it in mind, nonetheless.

Care and technology are no opposites (Mol et al., 2010). One is not warm and cuddly and the other free of attachments. One is not overwhelming and overreaching and the other efficient and contained. Technologies are not instrumental alone, they are part of action, of becoming. Technologies do not run in and of themselves, rather they "depend on care work. On people willing to adapt their tools to a specific situation while adapting the situation to the tools, on and on, endlessly tinkering" (Mol et al., 2010, p. 15).

Caring for digital security and identifying care in a corporate digital security practice provokes our understandings of care. It facilitates what is important to STS: critiquing moralities of blaming and judging and individual responsibility; forging relations rather than separating regions and drawing boundaries; facilitating collaboration instead of taboos. But it also encourages to further the concept of care to be sharpened and discussed.

Care does not scale up, but it refers to and dwells in the concrete situation. Yet, it is accountable and responsible for moral goods. By employing care in my stories of fragile computing, I respond to its current proliferation and hopefully to the vitality of discussions on the logic of care (Mol et al., 2010, p. 7). As scholars interested in care and digital security, we need to ask how the subject of care becomes concerned with more than isolated objects in contained settings; how to develop effective practices of care that scale up from tinkering with individual devices to address issues of infrastructure as they pertain to contemporary culture.

What emerges from such a perspective is more than the simple realization that digital security can be thought of as care, but the realization that such caring for digital security is conflicted, neglected, and troublesome. It conveys conflicting moralities and obligations. And it cannot be the intention to smooth them out, but rather to better understand the conflicts of digital security. It is this fluidity under distress that I find important to acknowledge for social theory.

Care work suggests a still understudied mode of the acting upon fragile objects (Denis & Pontille, 2011, 2015). Instead of describing stabilization work as ANT has done, care attends to invisible practices and continuous engagement to make technologies work. Digital security can be seen as complementary to existing studies on maintenance and repair that have emphasized the shadowy and underrepresented side of technologies. Instead of innovation and progress, these practices are oriented toward restoration or carrying on (Jackson, 2014). I have shown that, like other care work, digital security requires improvisation, bodily commitment, and long-term engagement. It requires stamina and staying with the tension between applying standardized solutions and adapting or exchanging them with local needs. That work is notoriously hard to account for, potentially endless, and invisible in formal plans and project management.

The previous chapter has introduced digital security in different forms, following patterns of relating (regions, networks, flows. This chapter has employed the lens of care to propose an emphasis on the nitty-gritty and morally charged work of doing digital security. Care fits fragile computing: it is not aimed at solving a shortcoming, but at making it a little better (or worse) (Danyi, 2022). This includes attending to matters of uncertainty on the grounds and modes of exploring and finding what secure means in a particular situation. Such care is fundamentally engaged in material actions while also conveying moral and political commitments.

For digital security, care points to the nitty-gritty work that is needed to achieve digital security in a way that is "good" for a particular situation. Digital security is often unsatisfying, imperfect, worrying. The chapter has introduced empirical stories where digital security work can be perceived as a matter of *care*: the object of continuous and collaborative tinkering where things are at stake.

References

Cohn, M. L. (2016). Convivial decay: Entangled lifetimes in a geriatric infrastructure. In *CSCW'16 Proceedings and companion of the ACM conference on computer-supported cooperative work and social computing*, San Francisco, CA, pp. 1511–1523.

Danyi, E. (2022). *Melancholy democracy: Politics beyond hope and despair.* Habilitation submitted, Institut für Soziologie, Goethe-Universität.

Denis, J., & Pontille, D. (2011). *Materiality, maintenance and fragility: The care of things*. SSRN. https://ssrn.com/abstract=1947255.

Denis, J. & Pontille, D. (2013). *Material ordering and the care of things*, CSI Working Papers Series 034, Centre de Sociologie de l'Innovation (CSI), Mines ParisTech. https://ideas.repec.org/p/emn/wpaper/034.html.

Denis, J., & Pontille, D. (2015). Material ordering and the care of things. *Science, Technology, and Human Values, 40*(3), 338–367. https://doi.org/10.1177/0162243914553129

Fisher, B., & Tronto, J. (1990). Toward a feminist theory of caring. In E. K. Abel & M. K. Nelson (Eds.), *Circles of care: Work and identity in women's lives* (pp. 35–62). State University of New York Press.

Gilligan, C. (1982). New maps of development: New visions of maturity. *The American Journal of Orthopsychiatry, 52*(2), 199–212. https://doi.org/10.1111/j.1939-0025.1982.tb02682.x

Haraway, D. J. (2016). *Staying with the trouble*. Duke University Press.
Heuts, F., & Mol, A. (2013). What is a good tomato? A case of valuing in practice. *Valuation Studies, 1*(2), 125–146. https://doi.org/10.3384/vs.2001-5992.1312125
Houston, L., Jackson, S. J., Rosner, D. K., Ahmed, S. I., Young, M., & Kang, L. (2016). Values in repair. In J. Kaye, A. Druin, C. Lampe, D. Morris, & J. P. Hourcade (Eds.), *#chi4good: CHI 2016*, San Jose, CA, May 7–12 (pp. 1403–1414).
Jack, M., & Jackson, S. J. (2016). Logistics as care and control. In J. Kaye, A. Druin, C. Lampe, D. Morris, & J. P. Hourcade (Eds.), *#chi4good: CHI 2016*. San Jose, CA, May 7–12 (pp. 2209–2219). Association for Computing Machinery Inc. (ACM).
Jackson, S. J. (2014). 11 Rethinking repair. In Gillespie, T., Boczkowksi, P. & Foot, K. A. (Eds.) *Media technologies: Essays on communication, materiality, and society* (pp. 221–239). https://doi.org/10.7551/mitpress/9042.003.0015.
Kocksch, L., Korn, M., Poller, A., & Wagenknecht, S. (2018). Caring for IT Security: Accountabilities, moralities, and oscillations in IT security practices. *Proceedings of the ACM on Human-Computer Interaction, 2*(CSCW), Article 92. https://doi.org/10.1145/3274361.
Latimer, J. (2000). Socialising disease: Medical categories and inclusion of the aged. *The Sociological Review, 48*(3), 383–407. https://doi.org/10.1111/1467-954X.00222
Law, J., & Lin, W. (2010). Cultivating disconcertment. *The Sociological Review, 58*(2_suppl), 135–153. https://doi.org/10.1111/j.1467-954X.2011.01966.x
Martin, A., Myers, N., & Viseu, A. (2015). The politics of care in technoscience. *Social studies of science, 45*(5), 625–641. https://doi.org/10.1177/0306312715602073
Mol, A. (2003). *The body multiple*. Duke University Press.
Mol, A. (2006). Proving or improving: On health care research as a form of self-reflection. *Qualitative health research, 16*(3), 405–414. https://doi.org/10.1177/1049732305285856
Mol, A. (2008). *The logic of care*. Routledge.
Mol, A., Moser, I., & Pols, J. (2010). Care: Putting practice into theory. In *Care in practice* (pp. 7–26). https://doi.org/10.14361/transcript.9783839414477.7
Nowviskie, B. (2015). Digital humanities in the anthropocene. *Digital scholarship in the humanities, 30*(suppl_1), i4–i15.
Pinel, C., Prainsack, B., & McKevitt, C. (2020). Caring for data. Value creation in a data-intensive research laboratory. *Social Studies of Science, 50*(2), 175–197. https://doi.org/10.1177/0306312720906567

Puig de la Bellacasa, M. (2011). Matters of care in technoscience: Assembling neglected things. *Social Studies of Science, 41*(1), 85–106.

Puig de La Bellacasa, M. (2012). 'Nothing comes without its world': Thinking with care. *The Sociological Review, 60*(2), 197–216. https://doi.org/10.1111/j.1467-954X.2012.02070.x

Puig de La Bellacasa, M. (2017). *Matters of care: Speculative ethics in more than human worlds*. University of Minnesota Press. (Posthumanities, v.41).

Singleton, V. (2010). Good farming control or care? In A. Mol, I. Moser, & J. Pols (Eds.), *Care in practice. On tinkering in clinics, homes and farms* (pp. 235–256). Transcript.

Toombs, A. L., Bardzell, S., & Bardzell, J. (2015). The proper care and feeding of hackerspaces. In *Proceedings of the 33rd annual ACM conference on human factors in computing systems - CHI '15* (pp. 629–638). ACM Press. doi:https://doi.org/10.1145/2702123.2702284.

Tronto, J. C. (1998). An ethic of care. *Generations: Journal of the American Society on Aging, 22*(3), 15–20.

Verran, H. (2001). *Science and an African logic* (2nd ed.). University of Chicago Press.

4

Training

I balance a black notebook on my lap, my view alternates between an oversized clock and a table in the middle of the room. I note down the exact time to the second and capture to my best ability all that is said, though, only fragments appear in my notebook: "How much money was lost?". "We must cut the network immediately." "We must come up with a shift system to respond to inquiries." This is a digital security crisis unfolding. On a 55-inch screen on the other side of the room a document is filled hastily with bullet points: "cut", "appoint", "communicate". Between the screen, the clock and me, a group of first responders sits around a round table in a room designed for this purpose. Similar rooms have large screens with security footage, this one does not. Disregarding the observers in the four corners of the room with their notebooks and clicking laptop keyboards, they were called into the room to coordinate a response to the cybersecurity incident that was reported just 30 minutes ago.

In early September 2017, I was invited to participate in an unannounced training exercise of Covy's Computer Emergency Response Team (CERT). This chapter describes the practical process of the training—preparing, assembling, and evaluating—attending in particular to how urgency is being crafted through the work of organizers. Digital security comes in the form of establishing an organizational "stand-by"

L. Kocksch, *Fragile Computing*, https://doi.org/10.1007/978-981-99-9807-4_4

(Taylor, 2021), acquiring competencies to respond when an incident breaks out.

This chapter unfolds in a sequence of events how a digital security emergency is prepared among a CERTeam. The exercise is observed from preparing the training to evaluating it by the team of organizers. During the exercise, digital security is created in the form of an event—a sudden incident that requires responses. The event enables through a few rounds of negotiation to arrive at a specific ontology of digital security that can be handled. The event form of digital security emergencies, however, is contrasted with the exercise of building long-term relations, acquiring routines and a repertoire of actions. While digital security is often portrayed as an event—a time-capsulated happening—exercises like these suggest digital security as long-term relations. Instead of rhythms of leak and fix, digital security is slow; instead of alarmism, it is mundane. And instead of striving to solutions that can be evaluated, it aims to handle fragility. The chapter ends with three implications discussing the role of trainings in digital security pedagogy and drawing together the previous chapters (Testing, Tinkering and Training).

Georg's work in the previous chapter has shown that digital security work is not one-off, but in practice stretches across longer periods, changing its shape and target, shifting between organizational boundaries. But digital security does not occur in the form of in-situ adaptation and tinkering. There are practices in place that attempt to hold digital security together in the long term. As is expected of a fragile form this is done less by keeping everything in place but rather by adaptation, shifting accountability, and anticipation and fuzzy futures. This chapter contributes insights into how preparing performs digital security. It explains how the practices of preparing enact digital security in a particular way. In the logic of preparing, digital security is characterized as an imminent collapse that is just around the corner, which proposes modes of action that are anticipatory, waiting, and being on stand-by.

Preparation is part of living with insecure technologies. It is driven not by the idea of prevention, but by living with the realization that a cyber incident can happen at any time. Training, thus, contributes to an alternative logic of enduring insecure technologies rather attempting to fix or solve them. However, it also takes a managerial perspective on such endurance: trainings are created to hone trainees' abilities in responding "correctly". Unfolding this tension—between a practice for fragile times—and the creation of assurance belonging to an age of control and progress—motivates this chapter.

Unexpected

Covy implemented a Computer Emergency Response Team a few months before my study and had already conducted three exercises before the one described here. I had spent several months in the company before the exercise and even though I had participated in many more spontaneous and ad-hoc moments of enacting digital security during coding, in meetings, or during testing, the exercise occurred to me as a pristine moment to observe how digital security comes into being. Wouldn't it be during trainings that that one version would be preferred over others?

On a sunny Friday morning in September, I arrive at Covy's campus, it is one of the first cold mornings of the year. Everything is calm and the only noise is faint quacking from the many ducks that inhabit the ponds around the campus. My gatekeeper Siegfried has invited me to participate in the exercise he is leading as the head of CERTeam. What better opportunity to observe cybersecurity than in a computer emergency response exercise?! When I arrive in the IT Security department, Siegfried walks up and down the office in a hurry. He briefly says "hi" and directs me to an empty desk and chair. He will be back in a second, he explains before running off again. Covy's cybersecurity department has an open office space with about 30 desks. Brick walls make sure no passers-by can glimpse people's screens. The desk I sit at appears empty; a standard swivel chair, document drawer, and an empty desk. A glass wall with strengthened milky glass patterns blocks me from seeing the person on the other side. I only hear them on a phone call, a shadow idly moving behind the milky glass.

Siegfried returns with DIN A-4 copies in his hand. He passes one to the silhouette sitting at the opposing desk, disrupting their telephone call. Siegfried puts his finger to his mouth, suggesting the person should not tell anyone about the piece of paper they had just received. Siegfried smiles conspiratorially and the other keeps silent for a moment and nods. Siegfried approaches me and hands me a piece of paper as well. Siegfried explains: "This is an emergency response exercise; it should come unannounced!". He grins.

I start reading the paper; it is titled "CERT table-top" and some names are listed in the header. Siegfried interrupts by sitting down next to me on the filing cabinet and taking the paper out of my hand. "Be careful how you hold it!" he warns. I am uncertain if he is serious or attempting to build up the atmosphere for me. Being handed the paper again, I feel like I am holding the correct answers to an unannounced test. "Security accidents could happen anytime, even on a regular Friday!", Siegfried adds, smiling.

In the version of the training, security is performed as something that arrives unexpectedly and without a warning. It requires a set of techniques to mimic this unexpectedness before and during the exercise. In the security department, Siegfried asks for secrecy and does so silently by putting his finger to his lips. He warns me to hold the paper in a way that no one can peek. The architecture of the security department aids in his attempt to keep the imagined incident secret, allowing nobody to hear or see what we are discussing. Siegfried is excited in anticipation of the exercise.

When in stand-by, becoming insecure is imminent, a security incident can happen anytime, even on a Friday. And it must be responded to immediately.

Siegfried begins by filling us in on our task. He explains that he is usually the head of the exercise. The challenge is to "step back from that and act as an observer". He warns us to not interfere or help, and act as if we were not present. This particularly applies to him because he is also officially a part of the usual line of communication as a security manager at Covy. Without further discussion of his strange position today, Siegfried explains what we should do during the exercise: "This exercise is about how the colleagues react to a case in the bank". Everyone seems to listen carefully, but I am the only one taking notes. The observational team should evaluate if "the participants verify the information and […] come before the situation". He begins to explain the concrete case the group is working on today: "The problem of the scenario is not in our own IT but comes from an external IT service." Siegfried explains the relevance of such a case by stating that "the regulatory agencies are very eager to make sure these dependencies are taken into account". Siegfried has told me earlier that this is the first time the team has had to involve a partner infrastructure. There are a lot of intersections between the bank and insurance and thus the exercise participants have to be cautious about any cascading effects.

Siegfried describes here the particular challenge he designed for the participants: The case did not come up in the internal infrastructure but at one of the partners. He argues that regulators want cascading effects between systems taken into account. On the other hand, he ensures here that observers know their tasks. They cannot interfere or interact with participants but must remain in the background.

The ritual play or "rehearsal" as discussed by Claudia Aradau (2010) in this case takes the shape of maintaining unexpectedness and mimicking a probable yet challenging case. This is specific for the exercise in this example as public health exercises are often announced and organized with reliable planning; the enactment then resolves around the actual practices during the training. The exercise here starts well before participants arrive. Mimicking the unexpectedness, characteristic for the preparing mode of digital security, begins when designing the scenario and making sure no one sees it and with warning me to hold the paper correctly.

An exercise must seem "real", which encourages "players to take the exercise seriously" (Adey & Anderson, 2012, p. 107). Siegfried's secrecy reflects his effort to make the incident seem realistic and tangible. Creating a proper equivalent to an "actual" case thus entails material work as well as crafting attachment between participants and the exercise (ibid.), and in this case, between observers and the exercise. Creating an atmosphere of surprise and secrecy for me and the colleagues in the departments is part of how Siegfried performs the suddenness of a possible security incident.

His efforts pay off when he describes the case and receives immediate responses:

Siegfried speaks calm and with gravitas. He explains the scenario looking at his piece of paper. Others follow his lead and read the scenario while Siegfried explains it. "Someone reported that there was an intruder in the connection from Covy to the bank". "The manipulation came from their own network". Someone replies sarcastically "Oh, great!", laughing ironically.

The sarcastic statement shows the participant's willingness to perform the incident. The scenario is plausible as the reaction indicates. In the mode of preparedness, a digital security incident like this could happen every day: it does not take much imagination by the participants to be intrigued by the case. The sarcastic undertone gives it a moral connotation: A manipulation from their own network is not "great" but means that an employee either has malicious intent or has made a mistake. This is serious as employees are trusted, and thus can cause greater harm. The training invites some moral judgment as to where the incident could have happened, but this is not what the group focuses on in what follows. Rather than tracing out where the attack came from, they shift toward organizing their collaboration and coordinating what needs to be done.

So far, I have pointed to the way Siegfried mimics the unexpectedness with which digital security is enacted in the mode of preparing. He creates an atmosphere of secrecy and equips the CERTeam room in a particular way to be able to accommodate both participants and observers. Finally, he prepares the observers for their role in the process, becoming active members of reaching preparedness by providing evaluations and an outside view of the training. In the next step, I will highlight the feeling of urgency that is introduced to craft the exercise in a way that applies real-time contingencies.

A central factor in mimicking an emergency is making the training "real-time". Urgency is reached with a rapid reach call infrastructure. All responders receive an automated phone call on their cell phones that instructs them to come to the CERTeam room "immediately". It also informs them that this is a training.

The urgency of the situation has an impact on my fieldnotes that have very accurate time stamps from this point on. My fieldnotes read like the protocol of a controlled chemical experiment, reacting in minute detail to every change in the room.

Siegfried sends the Rapid Reach call at exactly 10:20. His head turns back and forth between the large clock on the front wall of the room and his phone and the other observers. "It should take them no longer than 10 minutes to arrive. This is an emergency", he explains his tension to me. The rapid reach system is reserved for police and fire department emergencies, but if they do not need it, CERTeams can use it. Siegfried explains: "If nothing gets in the way, everyone should get their call soon". He looks at his phone. Nothing so far.

"If someone cannot come, they have to respond and say they are not available. This is important so the replacement will be informed," the CERTeam member explains to me while she opens a large table on her computer screen. Siegfried, the observer, and I stand around her laptop, our views glued to an Excel sheet. "This is an overview of the responses the people sent". Seconds later the first fields change their colour from grey to yellow and some soon after to green or red. One field remains grey.

The table allows about 20 phone numbers to be monitored that are listed on the very left of the sheet. Turning from grey to yellow means that people have received the phone call. It is not clear what their response is. Only when the fields turn from yellow to green or red, do the observer and Siegfried know whether this person will come, or if their substitute is informed. The field that remains gray is cause for concern for Siegfried and the observer. The observer begins to speculate: she explains that the person getting this call could have no reception as he commutes by subway and there are some tunnels with bad reception. The data is not personalized in the sheet, and I do not recognize the numbers, but she does. After a brief conversation with Siegfried in which they realize that this person does not have a replacement, it is decided to just wait a little longer for a response. Slowly more fields in the table turn from yellow to green. The observer explains to me that now people respond whether they are coming or not. Some fields turn red. The red fields are not discussed anymore as it is quickly identified that those "fields" have substitutes. Siegfried and the observer are satisfied by the responses which indicate that about 12 people will come to the exercise. The sheet remains open but is no longer consulted when people arrive.

I want to spend a bit more time on the rapid reach system because it provides a central device by which the mode of preparing crafts an incident: It operates as a constant reminder of the responder's permanent position on "stand-by" for an incident.

Siegfried suddenly becomes frantic: "Do I have to respond that I will come or not?" One of the other observers calms him down "You are already here!" The response system introduces tension. Siegfried's phone rings. He answers the call and puts it on speaker. An automated voice announces: "This an emergency. Please assemble in the CERT room immediately." "The system demands a reply", Siegfried explains in response. And the voice adds: "Press one if you are coming, press three if you are not available". Siegfried presses three and explains that his replacement has to be informed. When he hangs up, his tension lessens; "Everything worked out!"

Not all responses unambiguously show that both the rapid reach system worked, and that people are coming. Siegfried must decide whether he fulfills the system's requirement or not. He decides to respond that he is not available, forwarding the call to his substitute, which is necessary because he will have to lead the session today. Removing Siegfried from the training causes some difficulties not only for the responders but also for the system.

People's reaction times to the emergency call and their arrival times at the CERTeam room are meticulously controlled by the system and noted by Siegfried. People must show up within ten minutes. The rapid reach call and Excel sheet contribute to the design of an urgent case: Every color shift is recognized and the system creates a designated time stamp. Reaching out to people on their mobile phones (company issued) additionally creates their position of "stand-by". The phones are a constant link to being called into action. The rapid reach system furthermore contributes to the mimicking of the case by employing the rapid reach infrastructure that is also used for "serious" incidents.

After the call goes through and Siegfried has made his own choice of how to answer, he is relieved that things have worked out contributing to my perception of his anxiety about the rapid reach infrastructure. The rapid reach system and Excel sheet demonstrate an infrastructure that is in place that can be activated at all times (at least when there is no other incident or training). Not only is the infrastructure on "stand-by" awaiting to be used, but the responders are also in a constant state of waiting for the call to ask them to convene in the CERTeam room. "Standby", A.C.E. Taylor (2021) writes, arises as "more-than-material security practice, generating anticipative affects", e.g., working on high alert (p. 62). The real-time character of the training engages with this stand-by, allowing responses (and lack thereof) to be immediately linked with employees' practices and organizational options (e.g., finding a substitute).

The real-time character of the training continues as people slowly shuffle in. I invite you to read a longer extract from my fieldnotes transpiring the tension in the room waiting for participants to arrive.

> *Just one moment later, Siegfried is back in a state of tension, and the room becomes silent again. Everyone takes their observational positions. I sit down in my place in the back-left corner and my view rotates between Siegfried, the door, and the giant clock that is visible from almost every point in the room. Siegfried stands at the front of the room, anxiously folding the piece of paper with the scenario in his hands. His view is focused on the door, from time to time he looks at his watch. At 10:25, the first participant arrives, immediately relieving some of the tension in the room. Siegfried greets him by handing him an A4 piece of paper with the scenario description. Neither the participant nor Siegfried makes it explicit that this is an exercise and not an actual emergency.*
>
> *I assumed my position in the back. The first participant chooses a department tag from the centre table and sits down. It seems he does not have a fixed spot and just chooses one spontaneously. He and Siegfried chat briefly; they call each other by their first names.*[1] *Warning that it will take about an hour, Siegfried adds that he hopes he has the time. The other responds: "I always have time"; both laugh.*
>
> *Another participant arrives at 10:28. Two arrive at 10:30, one of them immediately says without addressing anyone in particular, that she "really" does not have more than an hour. After a brief discussion with Siegfried, she calls her substitution (manually) and asks him to come. Within minutes he arrives and explains he just has to send an email to cancel another meeting. Nobody else seems to need to cancel anything for this surprise meeting. After all, it's Friday, I think.*
>
> *Three more arrive at 10:31. The table slowly fills, it's quiet in the room. The participants who have already arrived get the scenario sheet from Siegfried and*

[1] While it was very common to use the German "Du" at Powino, in Covy, most people used the more formal "Sie". During the training, however, the atmosphere was more colloquial and relaxed and people seemed to know each other. The "Du" was more appropriate.

start reading silently or sitting at their laptops, there are no less than usual casual. People are curious about the scenario and most give it at least a short read-through. I recognize Sven, whom I met before at a security department meeting. He chooses the leader position without hesitation. Everyone receives a piece of paper, picks a department tag, and sits down on the table. At 10:33 one more person arrives and after she sits down, she is reminded by several people simultaneously that she has forgotten her department tag. They reach over to give it to her. The tags appear important even though I never see them being mobilized during the exercise. After everyone arrives, 11 people are sitting around the table—almost all the seats are filled. The organization seems to be appropriately represented as all tags are now taken; there was no dispute over who takes what tag and who leads the session. The hierarchy appears clear, everyone knows what to do.

The episode also depicts the atmosphere of the training as inviting joking and play as well as enacting security as an emergency that needs an urgent response. The crafting of a digital security incident as being urgent and requiring instant response affects the way people behave during the training: they arrive swiftly. The relationship between Siegfried and the participants is relaxed and they joke around before the training begins. But Siegfried is also very competent in making the situation strict and serious again, purposefully performing the incident as a potentially catastrophic event. In that way, the atmosphere of the training directly performs the digital security incident and creates powerful speculation about the future that affects the present.

It is fitting that what follows is tightly scheduled, limiting participants' discussion time to one hour. This contributes on the one hand to the real-time character and theatrical performance of the exercise and creates certain anticipation about how future incidents evolve (they need fast responses).

The session starts with Siegfried standing at the empty front of the room with only one piece of paper left in his hand. After waiting a short while for all chatter to die down, he starts talking slowly and calmly, welcomes everyone, and thanks everyone for coming so fast. After everyone had been typing on their laptops, the room's attention is now focused on Siegfried. Siegfried had been in a good mood and joking and laughing, but now his mood changes. He becomes serious and chooses every word wisely. "You have one hour to fulfil this task." He explains that in this hour they should find appropriate responses and report them to the "crisis management council". This is a council of Covy managers that approve and execute high-level decisions in the company. He explains that the leader of the training, he uses his first name, will report to the council after 45 minutes. The council is played by the other participants.

The training is arranged in two parts. The first is reserved for coordination among the responders, the second is for reporting to the crisis management board. This performs the incident as requiring a quick response, leaving little time for negotiation or discussions. The goal is a clear suggestion for the council. During the training, it is often repeated that "the council demands answers", adding to the imagination of the case. The next section describes the collaboration and coordination work done during the 45 minutes before the crisis board meeting.

Collaboration and Coordination

The training started with Siegfried handing over to Sven who leads the discussion. What follows is a partially chaotic (to me) back and forth of planning the next steps and attempting to verify what has happened. I perceive this as collaboration and coordination work. Mol notices that to counteract a case—of atherosclerosis or digital security—one must "coordinate" possible ontologies. This resonates, yet again, that there is no objective truth one can refer back to, but rather tediously assembled multiple objectivities.[2] While clinical and pathological diagnosis, in Mol's example, can be made in different practices, and in fact, atherosclerosis can be different things in these practices (a problem walking, a cold leg,

[2] Crucially, Mol and others interested in multiple ontologies are not after deconstructing reality, as was suggested by critique of STS in the early 1990s. Scientific facts then, digital security now, is not "socially constructed", meaning a political, cultural, or social fabrication. Quite to the contrary, tracing how ontologies are practically constituted involves various observable practices and objects. If anything, ontologies become "more" rather than "less" factual through an unearthing of their material-semiotic equipment. Here sits a central argument also made by Haraway (1988) in suggesting that "situated knowledge" is not a specific perspective *on* reality (and thus "less" objective than the god's eye view), but rather a partial and political specificity; and if anything, more objective than generalist views could ever be.

a particular blood cell count, etc.), when a decision is due on how to treat a patient, a single version must be decided on. Mol writes that it then requires "[a] patchwork singularity, the disease-to-be-treated of a specific patient. A composite reality that is also a judgment about what to do" (Mol, 2003, p. 72). By responding, a case or ontology is settled. During the exercise, a particular version of digital security becomes responded to. On the one hand, tasks are allocated to various representatives from different departments, hereby crafting a single version of security through collaboration. On the other hand, the group evaluates itself in a second step, thereby producing a version of security through coordination.

> *Sven leads the discussion: He holds the piece of paper in this hand and waves it to make a cracking noise. He begins: "Money was lost. This is confirmed, right?" He has opened a one-note document on his laptop that is connected to the big screen at the head of the table. He puts the piece of paper beside his laptop and starts typing. He looks at the scenario several times again during the exercise. Sven continues typing while someone raises his voice: "Is it confirmed?" They are suspicious and attempt to verify it. I draw a little star in my notebook confirming that they excelled in doing what Siegfried has asked them: "Do validation work".*
>
> *"Did it actually happen?", someone asks. They begin speaking of possible causes and precautions to take. In reaction to the question whether it was verified, Sven changes his notes in the documentation. A participant reads out loud from the scenario: "'An employee claims there was a manipulation from their own network.' Is it verified?"*
>
> *"We have to verify it," someone realizes.*

Instead of listing clearly that they would verify the case and define what exactly has happened, they continue by defining tasks, such as organizing a shift plan and naming substitutes. Comparable to doctors and pathologists in Mol's case who, by deciding to conduct a certain operation or therapy, also "verify" what sort of atherosclerosis the patient suffers from, participants in the training define the case by defining measures: Creating a mailing list and hiring technicians points to malicious code that may travel through network connections. By defining a shift plan, a response surprising to me, security becomes an organizational response, pointing to an ontology of security as a lack of coordination and work force.

The question of "did it *actually* happen" reoccurs several times in the discussion and is usually answered by various measures to take. For one participant, the question of how the incident was identified is what is important: "if it appeared at a routine check we can forward it to Siegfried and ask if he has heard of it". Occasionally, they rehearse what Sven pointed out in the beginning: "money was lost" which renders security a matter of draining assets.

Sven eagerly takes notes of the discussion in a document opened on a larger screen on the top of the table cycle. I cannot read the notes from my observational position, but Sven has a hard time finding the right "spot" to start the notes from this training. Other trainings have been documented here as well. By noting down hypothetical tasks they plan to fulfill after the meeting, the security case emerges. It takes its place in line with the old exercise notes. It translates the situation of the exercise and the discussion at the table from the scenario into a resource for future emergencies.

"Verification" was not done before measures were decided but throughout the process. The definition of the case and its measures are closely interwoven. The measures to take and the verification of the case are not chronological but interwoven, whereas certain measures cause different ontologies of the case. For example, Siegfried is only contacted for specific forms of an incident, but if they contact him, the response is clear, i.e., that Siegfried would enter the incident in his taxonomy and report it further.

It is important to note that verification (or nonverification) is not the sole ontological question here. Rather obviously, "verification" is not a binary issue but refers to methods of inspecting the case and measures to take. I argued above that the measures taken, in turn, shape the ontology of the incident, that is not solely *if* the incident is verified but *what* it is. This refers back to Mol's (2003) suggestion that when counteracting or treating a single patient a single ontology is fabricated.

One measure-ontology takes more time to discuss: Someone suggests disconnecting the network from Covy to the manipulated area. This, in turn, would mean halting payments to customers. "Then", he explains, "we have time to do some forensics to find out where the problem originated". In his suggestion, the ontology of the case is not immediately bound up with the response, but taking measures is followed by adding some forensic work, which means running an analysis script to investigate where a bug has occurred in the system. Rather than deciding during the meeting, his suggestion is to postpone and run more tests. Nevertheless, deciding to cut the network is drastic and has significant consequences. His idea is followed by several questions, e.g., if the customer lost money, how to communicate to customers, what payments would be affected, etc. Sven suggests:

> *"As long as we do not know what happened, we should close down all cash withdrawals and inform people of the consequences". "If there are really wrong payments happening", another adds, "we must stop all money transfers and inform the board". But "First we must inform the bank", Sven observes, "and this is the issue, we do not know if they can handle it". He decides: "We clip all connections and inform the bank". "First clip all interfaces between Covy and the bank, talk to the bank that no more payments are made". He continues while typing in the document: "We must develop a standard phrase why withdrawals are not possible at the moment" Sven makes sure: "Does the customer even know there is an issue—I don't use my online banking every day". "But I would be very disconcerted if my online banking does not work!", someone contends. "It just has to say technical difficulties", another one suggests. "and the call centre has to be informed". Another one suggests "Control social media and communication to the departments".*

During these coordination efforts, the group defines what measures to take when the incident is "really" wrong payments that are being issued. In this regard, once the connection is cut, the case becomes a different ontology—now it is a lack of service for customers, which, in turn, requires more measures to be taken. The group decides to suggest this procedure, which is later connotated as quite "drastic".

A second line of coordinating a measures/ontology is by referring to external entities that require a sound response from the group. The BaFin—Germany's financial observation office—is informed in cases of digital security incidents. Someone realizes that during the exercise: "The BaFin will ask what is going on". One asks: "Did they manipulate the data and what is it exactly, a Trojan horse or an employee". "We have no report obligation towards BaFin, if we filed charges, we would not have a complete story to charge someone on".

Someone again refers back to the way the incident was found in the first place: "If it came out in a routine check, then our measures work". If it was a "Trojan horse", "the story is different". When it was "an employee", it is different again. It is critical to him to mention that if the incident came out in a "routine check", the existing measures work—which helps them establish that tests do "catch" incidents. This attempts to approve one way of defining incidents: when they are caught by the security controls.

Toward the end of the session, someone warns that it is unclear when to inform the departments and that it is a rather harsh response to cancel all payments. *"Particularly, if the scenario is in the network, the bank is also in the network, so clipping our connection does not work. We have to identify the problematic connection"*. He suggests that *"we are too deep into defining measures and there is just so much that we do, and the rest does the crisis management"*. *"We can only discuss internally"*, he points his finger at Siegfried who smiles widely, *"We can only validate and analyse where and when something has happened and if there are any further manipulations"*.

I have argued there were three lines of coordinating the ontology of digital security here: Sven's first statement responds to the first line: an incident is when money was lost. This is countered by cutting the connection to the bank, stopping all payments effectively. By doing so, the case gets defined: an incident occurs when money was lost. A second line of coordinating ontology discussed here is an incident that must be reported to the BaFin. So, one could rely on external definitions of a case to identify if this incident is a digital security case. The third line is looking more closely at the measures in place and doing a "history" of the incident in order to define it. An incident has an incident "biography" in the system. The group decides for the first one, and after this point, the case is defined as "money was lost".

Crucially, the enactment of preparing necessitates coordinating ontological singularity, which was done above. I was most fascinated by the coordination of the case through losing assets, although, it was not very surprising, and Siegfried and the other observers took it as a given (and appraised it as successful "verification" in the evaluation later on). What this points to is that the preparing mode of digital security produces digital security in a particular form: as a single "incident", unexpected, something that needs an urgent response, and here, as one thing that must be coordinated, and can be responded to. At this point, there is no going back and choosing a different path of counteracting the case (let alone not perceiving it as a relevant incident at all).

According to computer scientist Adam Shostack (2014), three dominant ways of defining digital security issues exist in the area of threat modeling: "asset-centric", "attacker-centric", or "system-centric". He writes:

> Threat modeling is a fancy name for something we all do instinctively. If I asked you to threatmodel your house, you might start by thinking about the precious things within it: your family, heirlooms, photos, or perhaps your collection of signed movie posters. You might start thinking about the ways someone might break in, such as unlocked doors or open windows. And you might start thinking about the sorts of people who might break in, including neighborhood kids, professional burglars, drug addicts, perhaps a stalker, or someone trying to steal your Picasso original. (p. xxi)

Each entails a different ontology of what a digital security incident is: A threat to your family, a problem in your security system, a particular attacker. I am quoting this here not to rehearse a dominant form of how digital security is approached, but rather to demonstrate the ontological work that is part of the training. This points to other possible ways the responders could have managed the case. The group above follows a strict asset-centric logic. They do not wonder about the attacker or the specificities of the system.

But other forms might have been possible too: Thinking of a computer system less as a clear inside and outside but as a shifting and dynamic set of relations allows other questions to emerge. Poller et al. (2017) argue that large-scale organizations require other forms of threat modeling, e.g., thinking about relations and collaborations rather than strict entities (p. 70). It is crucial here that this does not simply patch up the existing threat modeling techniques but proposes a shift in scope. This is not a widening of threat modeling to organizational questions but demands to look differently at threats altogether.

I am discussing threat modeling here so concisely to open this point up for a debate about what forms of security can be translated into threat models and which cannot, e.g., the idea of an enclosed entity. This allows the discussion of other models of security to be introduced.

During the coordination phase of the training, different versions of digital security do not stand in isolation but become coordinated and thereby the asset-driven one is dominant. Everything that loses money is an incident that can only be responded to by immediately disconnecting from the bank.

The coordination work during the first 45 minutes was to enact one ontology of digital security as asset-centric which can be encountered by closing the connection to the bank. This goes along with the publishing of a digital security incident that will be reported to authorities to inform them of possible cascading effects. After this decision was made, Siegfried slowly gets up from his observational position. His presence in the room has an immediate effect. He gives Sven some time to sort his notes in the document before opening the crisis management meeting that includes Sven reporting to the group of responders that are now playing the critical council.

Shifting positions prompts the participants to evaluate the decisions they had made together, turning the last part of the training into "self-evaluation".

Self-evaluation

After the first part of the exercise, Sven must report the group's findings to a council of crisis managers:

> *He begins "Company E has reported—and this is validated", he adds "that a security incident happened. We do not have any indication of what happened where. We have taken quite drastic measures, disconnecting entirely from the partner company's network. No payment may be executed at the moment to contain the possibility of abuse". Sven invites the council to comment. One of*

the former responders raises his voice: "What payments are affected?" "Not the employees' wages?". He insists on details: "Also no insurance sums, damage control, etc. from the insurance?" Sven ensures that such payments are not affected. And closes the discussion by saying: "The misuse has to be prevented to avert reputation damage. We do this over social media and communications". Sven answers, "We should", he corrects himself, "we must", he corrects himself again, "we will announce the IT emergency!" A brief silence fills the room before Siegfried ends the meeting, thus allowing the participants to relax.

Sven corrects his tone to sound more self-assured and confident in the last part of the field notes. During the mock crisis management meeting, other logics of digital security come up that are introduced by one of the previous responders. He mentions the importance of the relation to the bank, employees' wages, insurance payments, and how to communicate to customers. There are two aspects to consider here: the first is that as a situation of self-evaluation, the curious responder gives Sven a bit of a hard time, and mentions things he has not mentioned before, putting the decision to cut the connection to the bank under pressure. Secondly, Sven's final words that we "will" announce an IT emergency create a cut-off point where the coordination work has come to an end and a clear definition of the digital security incident has been found: it is an incident in the bank, hence Covy will cut the connection.

I want to connect the dots here a bit more by pointing out how becoming prepared during the exercise links to the coordination of a single ontology of digital security: Mol points out in opposition to knowing atherosclerosis, "interventions" entail a form of temporary closure: "Therapeutic interventions do not primarily yield facts; they are supposed to change the object with which they interact. [...] They enact an object by altering it" (Mol, 2003, p. 89) and adds that "*Enacting* disease" in this case "takes the form of *counteracting* it" (emphasis in original, ibid.). From the manipulation that is reported, a list of responses emerges that finally coordinates into the announcement of an IT emergency before the crisis management council. Here, enacting digital security takes the form of counteracting the loss of money.

By deciding to plan a radical disconnection from the bank, the incident is enacted as something that is the effect of a tight network of services. A manipulation in one place (e.g., Covy) can cause money to be falsely transferred to another place. It is important to cut the network because the incident is defined as having possible cascading effects. The reaction of finding a shift plan points to another aspect of the ontology that emerges here: If a shift plan is necessary that means that services will be dysfunctional for a longer time, pointing to the fact that the incident cannot be contained in a short temporal frame. Rather, participants anticipate that the incident is so far-reaching that it requires a shift plan.

What this points to is that the enactment of the incident is done in a form of counteracting it here. Adey and Anderson (2012) employ the term "enactment" in their paper to mean that a future event gets played out and needs attuning to the materiality of the chain of events in play. But they also point to a more Mol'ian understanding of enactment when noting: "actions not only stabilized the identity and characteristics of the objects […] but actively brought them to life" (p. 112). The training crafts a space in which one ontology of digital security gets coordinated and acted upon.

This crafting of space for ontological singularity shows how digital security is performed overall in the mode of preparing: An incident demanding a response, with little room to apply anything other than previously learned definitions.

One final remark on how preparing enacted digital security. Preparing is sociomaterially assembled: the emergency call, the room, the document, Siegfried, the observers, participants, etc. Preparing puts devices, people, and organizations in a position of "standing by" (Deville, 2021, p. 98). The devices are meant to "lure" participants into the case, suggesting that a "real" incident has occurred and must be responded to. As argued above, with this, preparing settles on a particular ontology of the case. Furthermore, it requires to be drawn in and attached to the case. Emphasizing the feeling of being drawn in and affected by the trained case, we can make an argument for the way preparing requires a form of "touch"—of an embodied experience of the case at hand.

This touch experience suggests that rather than a process of objective comparison between secure and insecure (as suggested by some testing activities, see Chap. 2), preparing digital security trains people and technologies to negotiate and collaborate to identify what both secure and insecure are in a given situation. Instead of holding the incident against a comparison chart—or identifying a cut-off point—the case is coordinated and negotiated. Preparing for digital security thus entails material devices, Siegfried's work of creating unexpectedness and surprise, as much as an embodied experience and coordination of "the" incident.

I have argued that the CERTeam training was characterized by its unexpectedness, mimicking, real-time coordination and collaboration efforts, and the socio-material creation of attachment. In the conclusion, I will suggest how the CERTeam training resembles and differs from other preparedness exercises (Implication 5); draw conclusions for digital security research (Implication 6); and link the chapter to the overall argument of developing conceptual approaches for digital security in this book (Implication 7).

Preparedness

Training for potential disasters can be located in a security strategy commonly referred to as "preparedness". The following section summarizes key arguments in the literature on preparedness in anthropology and STS.

Preparedness becomes necessary when disasters are inevitable yet evade exact statistical prediction, e.g., pandemic events (Collier & Lakoff, 2008; Lakoff, 2008). Preparedness efforts aim at building a range of capacities and competencies to mitigate potentially catastrophic events (Collier & Lakoff, 2008, p. 14). Preparing for disasters means becoming experienced with the material form, shape, and distributed effects of an incident, something that often irritates contained forms of organizing (Burton, 1994).

Preparedness performs a particular temporality in which "waiting" is actively crafted (Deville, 2021). It links possible future incidents to actions in the present. Being prepared means being on "stand-by" for the unexpected (Deville, 2021), which includes that at any moment in time actors must be ready to be re-activated (Kemmer et al., 2021). Waiting for disasters is considered an "active inactivity", being in a permanent state of possibility (ibid). Relatedly, preparedness efforts are ultimately always unfinished and ongoing: Despite best efforts of planners and trainers, anthropologist Andrew Lakoff (2008) argues, preparedness will never come to be conclusive, and actors will always be unsettled by the prospect of future surprise. Despite best efforts, one will never be completely prepared. Thus, preparedness offers a mode of action to grapple with events that will remain uncertain, while having the bleak understanding that preparations are fragile.

Lastly, anticipating the future for preparedness is not neutral but invites moral judgment and an ideal of a right response to an approaching disaster. Anticipation is not only an epistemic reorientation toward performing the future but entails moral imperatives (Knox, 2020). A moral injunction exists in preparing to act in accordance with the avoidance of future harm. Preparedness exercises in the public invite a moral economy of acting well today to avoid worse tomorrow (Adams et al., 2009, p. 249). Preparedness builds public sentiment of "good citizenship" applying social control (Baker & Ludwig, 2018). Natalie Baker and Lisa Grant Ludwig criticize preparedness models' moral undertone and suggest it must be reframed to better understand the behavior of citizens in times of crisis. They point out that previous approaches, although they have captured preparedness conceptually, know little of the concrete practices in which preparedness is achieved and contested (ibid.). They suggest looking closer into local preparedness practices to recognize how people react to possible disasters (Baker, 2014). In her ethnography of earthquake preparedness in Southern California, Baker suggests a form of "situated preparedness" to account for her findings of implicit preparedness strategies and local differences (ibid.). A common understanding of preparedness that citizens would panic in case of disaster and must, if anything, be seen as a source of chaos and additional risk has been refuted in similar studies, fostering interest in the question of how planning and improvisation interact in times of crisis. There is now increased attention to individuals' everyday lives, and their "informal preparedness activities" that stray from official guidelines (Heidenstrøm, 2019, p. 379).

Traditionally preparedness has applied to areas of military action (particularly during the Cold War years, cf. Collier & Lakoff, 2008); hazard adjustment (Lindell et al., 2009); pandemic preparedness (Lakoff, 2008; Wolf & Hall, 2018); natural catastrophes (Deville et al., 2014); terrorist attacks (Aradau, 2010); or emergency services, e.g., fire or floods (Deville,

2021); and their consequences for connected "vital security systems" (Collier & Lakoff, 2015).

Authors argue that preparedness accompanies a general shift toward "securitization" where present actions are evaluated according to their impact on an anticipated future (Buzan et al., 1998). Present actions thus become scrutinized through the lens of relating them to possible futures. Mobilizing the lens of preparedness for digital security research means foregrounding modes of dealing with inevitable disasters and making the future tangible in present actions.

Preparedness literature includes both conceptual considerations that observe a shift from exact calculation toward preparedness and mitigation of risks as governance strategies (Beck, 1992; Buzan et al., 1998; Aradau, 2010); as well as empirical perspectives on concrete practices of emergency training (Deville, 2021; Wolf & Hall, 2018); and practices of responding to disasters as compositions between implicit and explicit actions (Baker, 2014; Baker & Ludwig, 2018; Heidenstrøm, 2019).

Anthropologists Meike Wolf and Kevin Hall (2018) observe that "[while] most of [preparedness] accounts characterize [it] as a set of political rationales and concepts, still little is known about how preparedness is practiced locally, and which technologies and techniques are facilitated" (p. 487). In their analysis of pandemic preparedness practices in European metropolitan areas, they find preparedness to be composed of heterogeneous actors, forming what they call "cyborg preparedness", pointing to infrastructures, techniques, competencies, and human bodies that are equally engaged in becoming prepared (p. 495).

The present chapter has attempted to sensitize for the material and local achievement of preparedness for digital security in a similar way, pointing to how rooms, digital warning systems, Excel sheets, employees, and trainers become prepared and thereby enmeshing prepared futures with contemporary material settings (Deville et al., 2014). In my example, I propose that empirical, ethnographic inquiries into digital security emergency exercises are important to highlight how they operate as situations to train organizational collaboration and actively craft ontological singularity.

Exercises stipulate a type of theatrical performance, a "ritual play" that rehearses "habits of preparedness" (Aradau, 2010, p. 103). Exercises are intended to reduce the negative impacts of disasters and decrease social and physical consequences (Burton, 1994).

Exercises are common for emergency services like fire and police departments that train reactions to acute cases of flood, fire, or chemical exposure. During those exercises, the event is either simulated (with physical means like foam or smoke) or deliberately introduced. In the case of fire exercises, for example, objects are set on fire in a controlled environment, giving responders the chance to train their actions with a concrete experience of temperature and smoke development. Exercises attempt to bring the emergency into existence in the absence of an actual disaster but cannot always mimic a disaster in its physical unfolding (Deville, 2021).

The CERTeam creates the necessary infrastructures and routines to respond quickly and effectively, as well as analyze and report incidents to authorities. The team serves as "repository for incident information" as it functions as the organize point of contact for computer security emergencies. (Killcrece et al., 2003, p. 1). CERTeam are part of a wider network of security controls, where incident report fuel into other planning documents, such as risk management. CERTeams respond to an interest in standardizing digital security incident and emergency response that has grown since 1980 (Slayton & Clarke, 2020). Slayton and Clarke have argued that an analysis of CERTeams contributes to understanding the continuous work necessary to repair and maintain information infrastructures. This work has gone unnoticed for a long time (Slayton & Clarke, 2020, p. 4). They observe that most work in digital security has focused on the creation of new knowledge and technologies rather than on the maintenance of existing infrastructures (p. 5). Exploring CERTeam's work, therefore, counteracts the prevalent focus on innovation by emphasizing the work needed to maintain, repair, and prepare for the demise or breakdown of infrastructures. This focus on maintenance and endurance reiterates that the CERTeam training enacts a form of fragility, rather than innovation or progress, that is more common in prevention efforts (but not exclusively as previous chapters have argued).

Implications 5, 6, and 7

While preparing digital security relates both to the practicalities of imagining an incident and to questions of what "good" practices look like today. Preparing performs specific politics and invites distinct moralities through the enactment of a possible (seemingly inevitable) future (de Goede & Randalls, 2009). The practice of preparing enacts a specific temporality, and particularly, it implies a future that the preparing activities are oriented toward and that directs what is to be done in the present.

In that way, anticipating digital security futures is not a neutral task, but rather shows yet another specific enactment of digital security, that might partially (or fully) stand in conflict with other enactments. Crucially, perceiving of digital security as practices anticipating the future inquiries into what futures (in plural) are imagined, to be averted or desired.

Fragile computing in the practice of preparedness cannot be controlled, but only experienced; it "cannot be known or predicted; it can only be enacted" (Aradau, 2010, p. 2). This shift moves from aiming efforts at specific isolated problems to vaguer and more fuzzy potentialities. Preparing aims at making possible futures more tangible in the present, which delegates worrying about consequences in present action. Her note that such events can only be enacted speak to the specific performance of such events, requiring physical rehearsal and experiencing.

Three implications follow: On the one hand, the training allows to mobilize the lens of preparedness for the study of digital security, pointing to the tools, techniques, skills, and competencies contributing to preparing for a digital security incident. Preparedness training has been addressed in anthropological and STS research as crucial devices to "practice uncertainty" in participants and wider publics (Adey & Anderson, 2012), and across socio-technical assemblages (Wolf & Hall, 2018). However, scholars so far have not looked at practices of preparing for digital security incidents. Aiming at this gap, the chapter underlines what is particular about preparedness training for digital security emergencies.

Second, digital security is an organizational issue in practices of preparing; it requires collaboration and coordination between actors and includes specific enactments of the future. So far, CERTeam digital security trainings have received little attention in digital security research and even then, it was mainly from the perspective of transnational standardization (Slayton & Clarke, 2020).[3] This chapter looks closely at the concrete practices of the training, pointing to collaboration and communication as central tasks that are part of this enactment of digital security. This allows the scope of digital security research to be widened from technological and psychological perspectives on prevention toward how the preparing mode of digital security includes possible futures and governs present actions.

Other than user or developer exercises that are widely acknowledged in digital security research (exemplary: McCrohan et al., 2010; Caldwell, 2016), preparedness training includes a shifting from the prevention of an incident and targeting of individual responsibilities toward a coordinated response that spans across humans, technologies, and organizations. The chapter argues that digital security as preparing is heterogeneous: response infrastructures, timing, trainers, participants, communication, etc. Looking at the concrete material practice of the training allows inquiry into the set of tools, practices, conventions, and relations that contribute to digital security in this mode. Arguing that preparing for digital security performs a particular version of digital security, one can consequently ask what other tools, practices, conventions, and relations could include how digital security incidents are imagined and encountered differently.

Third, contributing to the notion of fragile computing, this chapter shows how digital security not only means facing digital security issues head-on in distributed and ongoing acts of care, but in training practices, it involves the crafting of anticipation of an impending breakdown. Digital security is not only enacted in preventive measures or ad-hoc adaptations (as a fluid form) but also in long-term speculations for possible futures. In the training, technologies are frail and susceptible, always on the verge of collapse. This turns digital security into something engaged in future relation-making, yet no less material and practical in

[3] Abbreviation CERT is also common to mean "Community Emergency Response Teams" aiming at a plethora of emergencies, not related to digital security (see for example Connolly, 2012).

the present. The exercise trains digital security as material work of anticipation and mimicking, which points to the need for coordination under pressure.

Implication 5: Digital Security as a Field of Preparedness

I have pointed to some characteristics of preparedness according to existing literature above: (1) Preparedness is necessary when disasters are inevitable but defy exact statistical calculation; (2) there is a need to actively perform disasters in their concrete material and time-sensitive dimensions; (3) preparedness efforts are notoriously unfinished and ongoing; and (4) preparedness efforts invite the moral judgment of "good" and "bad" actions with relation to the imagined future.

Existing literature on preparedness, however, has so far not attended to cyber-attacks. Some specifications should be considered what preparing for cyber incidents entails. The following section discusses how the CERTeam training above relates to each characterization and what the investigation of digital security preparedness trainings can contribute.

The exercise provides a case of preparedness training in the sense that it does not aim to prevent an incident or but enables practices of speculation (Lakoff, 2005). Attending to preparedness marks the enactment of digital security as contained "incidents" that can be more or less precisely located and will not be resolvable through technical fixes. The preparing mode of digital security puts an emphasis on its temporal dimension where incidents could already be ordained, written in the code, or maliciously introduced, but whose effects are yet to be seen. The exercise combines urgency, time pressure, and past mistakes with an unknown, but prepared, future to come. While surprise is a common theme in preparedness trainings (Aradau, 2010), I have shown above how the unexpectedness of the digital security incident is actively crafted—by maintaining secrecy and choosing a challenging case for participants (outside of their immediate reach).

Geographers Peter Adey and Ben Anderson (2012) differentiate between emergency exercises that have the goal of proving and legitimizing experts—often found in US public safety—and exercises that test material composites and enact capacity building—which Adey and Anderson identify as dominant in the UK's resilience strategy. The latter is rarely about unimaginable or *"unknown unknowns"* (one way Siegfried framed it), but about definite—albeit untamable—material unfurling of events. Designers of exercises of the latter types draw on existing or previous scenarios to build their cases, building step-by-step capacities in responders to coordinate and respond to a case. How scenarios are chosen is influenced by the experience of the designer. The training above falls into the latter category. Siegfried draws back on previous trainings and employs methods and ontologies taught to the participants before. In addition to exercises observed by Adey and Anderson, however, he also draws on his vast and intimate knowledge of the group of participants. He anticipates what might be difficult for them—the case of the external banking infrastructure—and when interpreting the results of the rapid reach system, knowledge of the participants helps to contextualize ambiguous data. Siegfried anticipates the cascading of emergencies with the way he designed the training, alluding to the idea that one event in a connected infrastructure can have vast effects on Covy's systems.

The fact that an emergency is trained that does not occur on its own infrastructure is, however, unusual for emergency response trainings and instead shows similarities to health emergency trainings where the origin of an incident is often unknown (Wolf & Hall, 2018). In contrast to emergency response trainings by fire departments or other first responders, the exercise felt quite removed from the actual incident. Nobody was sent out to fix code or talk to the colleague who had become the victim/potential accomplice. The scenario consists of a narrative story. It is explained that "something" was reported and that it was an *"inside manipulation"*. The incident is explained in detail: *"Someone called it in"*, *"the person remained anonymous"* (taken from the scenario description).

The digital security exercise, in turn, has similarities with emergency service exercises by mimicking the temporality of an incident that unfolds; it is "real-time" in the sense that an "actual" meeting of the CERT group in the case of a real emergency would not have more time to respond. And comparable to other emergency trainings, the CERTeam training included observers that would provide feedback and an evaluation of the process (Wolf & Hall, 2018).

Furthermore, similar to health emergency planning, the exercise was not designed to affect participants with audio-visual or other materials. Being comparable to other table-top exercises, the training intended to map organizational structures and lines of communication. Nevertheless, as shown, the ethnographic description of such trainings bears potential as trainings figure as important spaces of settling ontology and making moral claims.

As indicated in the literature, preparing invites moral judgment (Adams et al., 2009). This moral sentiment becomes very clear in the ironic *"Oh great"* when first being confronted with the scenario in which a colleague is blamed. I also made this point earlier in the moral accountability that Georg and Karolin face, but I want to reiterate it here, nevertheless. Enacting digital security entails moral injunctions, a form of political thinking that enlists individual action into a greater cause. The mode of preparing digital security marries material action with moral and political commitments. It provokes affective states, hope, and speculation as well as surprise, despair, and uncertainty (Adams et al., 2009, p. 249). During the training participants stayed relaxed and at ease. There was no big chaos or even panic. But the politics of the training became tangible, nonetheless. First, defining the case as "money was lost" has political and ontological consequences, as discussed above. Second, pointing to responsibilities in individual employees that (may or may not) have introduced malicious code, rehearses the moral register of blaming and judging. But the preparing mode does not stay with blaming and judging but identifies tactics to deal with insecurity. I will come back to that point of preparing as a strategy for dealing with fragility later.

While Lakoff (2008) states that preparedness efforts are continuous and inconclusive, trainings are temporally limited, and exceptional situations. Trainings cannot be continuous and ongoing; they are moments carefully crafted to perform incidents as time-sensitive events that require a clear and coordinated response. All while being aware that in everyday work, digital security may involve other temporalities (such as long-term engagement, waiting, postponing, etc.). The practice of preparing enacts digital security as a time-contained event. The training crafts urgency, not patience.

What must be viewed critically is whether the training was successful in aiding participants to imagine the unimaginable (Aradau, 2010). It seems participants rehearsed existing models of digital security for coordinating the case and making decisions on what measures to take. While the training then serves as a good place to investigate how prevalent models are coordinated and become effective, they do not provide a space of openness and creativity where unusual forms of digital security can be expected.

Finally, what has so far not been made explicit in the literature on emergency planning is how a distinct ontology of "the" disaster in question is coordinated in the course of the exercise. This gives reason to look closely at *what kinds* of securities are enacted in preparedness trainings, as these can be seen as arenas in which official agendas are stitched together. As Wolf and Hall (2018) note, it is crucial to see what kind of practices are moved to the center and which are left behind. In this case, the asset-centric logic has determined the training, and other possible ontologies/measures were dropped. I have drawn on Mol's (2003) understanding of "coordination" of various ontologies into one to emphasize the ontological dimension of emergency exercises. Crucially, measures of how to respond to the case and the ontology of the case are closely entwined.

Implication 6: Learning from the Training for Other Trainings

A version of digital security is enacted in the training through coordination and collaboration. It cannot be reduced to individual factors. This echoes a broader argument I make in this thesis. However, what becomes more explicit here is that there is much to learn from even the "official" versions of digital security that are enacted as almost theatrical engagements. So far, I have argued that it needs attention to practices that subvert and contest official digital security enactments for example through practicing fragile forms or acts of care. This chapter centers around the crafting of dominant account of security through events such as the training ceremony. Most noticeably, the coordination work goes along the lines of monetary loss or an "asset-centric" view of digital security; but others would have been possible. What if the responders did not assume the ontology of digital security as "money was lost", but had asked about why the case was reported in the first place and who blamed an employee? Or if they had focused on trust as a determining ontology to act upon.

What if the training had prepared for a more complex case, accounting for local practices of the ominous employee that was now blamed for the incident? What if the training was not aimed at a crisis management meeting but a group of customers? What if the case was decided to be benign and not to react at all? What if the reporting was simply a joke gone bad? What this exhorts is to think of other ontologies of "the incident" and widen the space of possibility to other lines of coordination other than assets (reputation, monetary loss, etc.). I have referred to existing different threat models to point out that thinking through and with different models about digital security is not completely new to digital security professionals. What it actually needs is adding additional models (which is the goal of this research).

One other possible enactment of the case was actually brought up during the evaluation session that I largely excluded from this chapter: The critique was voiced that participants did not include the banking division in their considerations. Among the observers were two from the partnering bank that had hoped to be included in the decision to cut the network. As a concrete critique to the training here, one can ask what the incident coordination and responses would have looked like had one included more "affected publics" (Marres, 2012), not just the bank but also customers, the blamed employee, etc. To STS scholar Noortje Marres, "affected publics" or "communities of affected" are those that are touched by yet rarely included in decisions about material devices and infrastructures. In her reading, material components, as much as gatherings like the training have the capacity of inviting affected groups in the conversation and allow for participation.

As Aradau (2010) notes, preparedness exercises do not create something new or urge participants to radically challenge prevalent notions (p. 4). It seems they rather manifest common lines of coordination and rehearse what participants have learned in prior trainings. I emphasize this here as I would encourage digital security researchers to think beyond such prevalent models and introduce more critical thinking in digital security. Some of the considerations in this book could be employed here. As I have noted, other earlier models of digital security could have been possible and could affect how digital security is practically attended to. Even though the preparing mode of digital security did offer a distinct new enactment of digital security, during the coordination work no existing modes were fundamentally challenged. Nevertheless, some aspects of the exercise can be mobilized for other educational settings in digital security.

In this respect, the training does have something to offer for other types of security trainings, e.g., awareness trainings. It is the core idea of table-top exercises to bring various actors together to find solutions that are accommodating of various areas of interest. Such collaborative scopes would be desirable for other types of security trainings, e.g., that aim at prevention. And first attempts of opening up the formats of such trainings have taken place (Williams & Coles-Kemp, 2019).

Furthermore, going beyond the scope of this training, in the design of user trainings, it is noticeable that preventing measures prevail, whereas it could be useful to also give users the means and methods to deal with incidents that have occurred. Preparing practices must not be closed off but could include users and other affected groups. They could better prepare how to deal with "smaller" issues, e.g., what if my account has been hacked, or my password has been guessed? Recommendations for users traditionally rest on individual solutions and mobilize moral judgments (change your password or else!), but do not take into account the complex practices of people and their respective stakes.

Inviting users into CERTeam trainings for example could add a perspective and offer ways of making incident response less asset- and more practice-driven. Preparedness trainings like this could also be employed among user groups, allowing strategies to be defined that are not tailored toward individual responsibility but collective action. Furthermore, preparing digital security with users would give both users and professionals the opportunity to learn more about the various stakes at play. In the example above, evaluation could have taken place on both "sides", among observers and participants, to capture what modes of action were considered but not fueled into the prevalent discussion.

The preparing enactment of digital security shows digital security fundamentally as a socio-technical and organizational question, whereas prevention trainings are usually tailored toward individuals (or regions, see Chap. 2). My suggestion is that users could also be invited to preparedness training to gain a better understanding of how to deal with digital security incidents (and potentially widen possible ontologies) as they perform a better understanding of dependencies and cascading effects. While this suggestion again targets individuals, it could be possible to translate other aspects of the preparedness training to prevention trainings. The preparedness training did not aim at individual responsibility, but organizational responsiveness, something that can be desirable for other trainings.

Implication 7: Preparedness as Dimension of Fragile Computing

The chapter points to how preparing enables becoming experienced with crisis and breakdown—or how preparing exercises "broken-world-thinking" (Jackson, 2014). Jackson argues that focusing on such computing practices that are tailored toward repair and maintenance shifts attention from innovation and progress to the daily, nitty-gritty, and underrepresented acts of living with brittle, outdated, or decaying technologies. While his focus is on electronic waste and the possible futures of material devices, taking digital security into account points to perennial breakdowns in fragile technological worlds.

Preparing for digital security incidents can be added to the investigations of actions like repairing, maintaining, caring, or recycling technologies as it emphasizes not innovation and progress but modes of dealing with what is there and what might sooner or later become frail. In this way, modes of preparing contribute to what I have termed "fragile computing": it includes practices aimed at living with uncertainty and inevitability of breakdown in technology cultures.

In this chapter, this fragile form is not held together with in-situ and continuous acts of care as it was in the previous chapter but by planning and anticipating the future. The precarity of this fragile form of digital security thus gains a temporal dimension.

This section combines some of the arguments made throughout the chapter with the topology of fragility that was introduced earlier. The training suggests a set of techniques that have been developed to attune to this fragility, e.g., the creation of an emergency response system, a responder network, and so on. While this could be seen as collective, or socio-material works, of becoming resilient; we might also take the "broken-world" perspective more seriously in asking what ontologies become possible if we can no longer avoid, but are already living with constant breakdowns. Preparedness handles breakdown in a different way than avoidance would.

Preparedness shifts attention to the "what-ifs" of contemporary societies, making possible futures impact the present (Adey & Anderson, 2011, p. 100). Preparedness thus contributes to the definition of a fragile pattern of action as it entails an enactment of what is constantly on the verge of collapse. In fragile forms, a breakdown evades exact statistical calculation and stays somewhat vague until it happens. The training crafts a moment where a singular case is managed, closed off from any other concerns that may be associated with digital security "in the wild" (Dourish et al., 2004).

This chapter adds the perspective of preparedness to my proposition of a fragile form of digital practices in which digital security gets actively orchestrated as technical, material, and organizational fragility of infrastructures that are unavoidable. Instead of assuming digital security as simply "evaluated" or "assessed", or as simply "fixable", this moves attention to what evaluations and assessments perform, what they exclude and conceal, and how they appoint responsibility, trust, and mistrust. The exercise emphasizes what is not fixable (anymore) but requires attention, communication, and collaboration. Nevertheless, the exercise sticks to strict asset-centric logic. Preparedness is perceived as "minimizing damage" in the form of money or reputation loss, thereby intervening in how digital security incidents in the future are imagined and performed.

Preparing is part of the topology of fragility in which one must assume that continuity could come to an end at any time without massive or abrupt changes in regions, networks, or fluid adaptations. In fragility, discontinuity is imminent; therefore, fragile spatialities include anticipation and speculation about possible futures. Adey and Anderson refer directly to Law and Mol's topologies, when saying that preparedness enacts a pattern that "flickers between actuality and potentiality, between local and global scales, collapsing the presence and absence of distant relations" (Adey & Anderson, 2012, 106). Preparedness responds to the notion of "fire", they concur:

> the qualities of fire describe the "multiple alterities conjoined to a centred presence"—an emergency characterized by sudden and unpredictable movements and changes of direction, excess, and surplus. This volatility is a product of the complexity of the vulnerable systems and the non-linear character of the emergencies they compose, which stretch across time and space. (Adey & Anderson, 2012, p. 106; Law & Mol, 2001, p. 615)

Fragility sits in these perennial oscillations of secure and insecure, in the spontaneous capacity of breakdown, and the stretching across temporal scales. I have pointed to this in the description of the training by paying particular attention to the temporality in which an incident unfolds, crafting urgency in a precarious situation.

I have argued in the previous chapters that fragile forms entail different modes of action; tactile knowledge and prolonged and patient caring attention. This chapter proposes that composing theaters of urgency and coordination also belong to living with fragile forms. When preparing, I aimed to portray, digital security is not calculated or predicted; it is not about engaging in a speculate future that could be fixed today, but rather about honing collective abilities in coordinating flickering objects. Instead of an attempt to reach beyond located practices, preparedness comes as a here-and-now practice. An obsession with how the training produces resilience (rather than avoidance) might however also distract from seeing the practical engagement with fragility that is part of this chapter. Striving to achieve resilience can be prone to forget our here-and-now mess. Albeit through an extraordinary training, preparing digital security is a case of "perfectly ordinary politics", not a rush toward latest solutions (Vinsel, 2023). Fragility is nothing special.

The exercise does not prevent the incident but lives with it and dwells in the troubles the incident can cause. It does not "fix" the issue but stays with it and attempts a possible "good" while knowing that things cannot be solved. But something feels odd finding that this exercise represents what Haraway has called "staying with the trouble" (2016): staying in a situation despite its troubles. One could conclude that the asset-centric ontology does include a specific corporate- and economic-driven way to perceive the incident, and one would be correct. Nevertheless, I find it worth noting that preparing for digital security shifts from solution- and prevention-centric modes of action toward patience and stand-by that stay with and dwell in what is broken. It causes certain discomfort to analyze such practices within corporate contexts, as anthropologists would look to expect them rather among activist groups or in ecological debates, but I want to let this unease be part of the theorizing attempted here.

Modes of preparing emphasize furthermore that things do not turn fragile at once, they become fragile over time and through the work of coordination. Preparing is necessary not only when the time comes but before that and continuously every day. There is not one big disaster that this training is planning for but also the continuous small disasters. I liked this training so much as it was an active performance of digital security that was rare on an everyday basis, and I did not experience one big hack. But what I did experience were smaller phishing attacks, and my anti-virus program certainly fought off some trojans or other critters that my computer picked up on the Internet.

This chapter has presented a preparing mode of digital security: Before, during, and after the training, digital security gets enacted as a singular case that can be tackled in isolation. Preparing work consists of establishing material devices and response infrastructure, as much as learning how to coordinate a case on the side of the participants. This coordination work is not solely "objective" but rather a process of negotiation and aligning possible responses with what has been reported.

I have argued that it is worth looking at how digital security is prepared, first, because it contributes to efforts in STS and anthropology to understand how preparing aligns possible futures with present action; and second, because investigating a preparedness training can inform how other digital security trainings are designed; and third, as preparing for digital security presents a form of living with and thinking with potential breakdowns that are characteristic for fragile computing.

Instead of in a preventative mode—practices aimed at identifying how vulnerable technology is—this chapter has approached digital security as a matter of preparing—practices aimed at dealing with fragility. Preparing invites a different mode of action: mimicking, anticipating, attaching oneself with an imagined future, which results in a different ontology of digital security: an incident that could already be there. This proposes a crucial specification to what I have termed "fragile computing": it entails practices that make a breakdown vivid in the present.

One more final remark should be made here. Participating in the exercise challenged my motivation as an ethnographer from "participating in order to write" (Emerson et al., 2011, p. 26) to participate in order to evaluate and intervene. My field notes were equally used to evaluate the exercise's outcome (another meeting that I have not included in this thesis). I want to note that my critical reading of the training as performance in imagining a case (not so much the anticipation of a "real" case) made my notes unsuitable for the evaluation. I identified the asset-driven logic of security as a problem in the evaluation meeting which was replied to as "rules of the game", which could not be overcome. The concern was not shared by the other observers. Nevertheless, my notes became more of a device through which the field reflected upon itself as I was introduced as a "professional observer" who could provide such reflective material. During the evaluation meeting, it appeared, however, that I had to comply with established ontologies in the field, e.g., the asset-driven tactics to convey a coherent evaluation of the workshop within the established criteria.

While collaboration between ethnographers and their field sites and alignment of ethnographers' goals with their fields is usually connotated very positively in STS, I felt conflicted and disconcerted to become a collaborator in this field. The tension between complicit and critique became particularly alive in this example as my ethnographic material was turned into a device of evaluation for the exercise. I have explored this issue at length elsewhere (Kocksch, 2020).

Being interested in the practicalities of doing digital security, the training offered a more theatrical and somewhat ceremonial enactment of digital security, where the practice of preparing for shaped what form of digital security was considered. The training crafted an exceptional situation in which digital security could be coordinated and tackled in isolation. Such moments are rare as things usually overlap, conflict, and are messier, as previous chapters have shown. Nevertheless, they are crucial as they operate as arenas in which ontologies become actively crafted and effective.

The next chapter underlines that digital security is rarely in the center of attention like that, yet still, is worth attending to. It points to the way digital security is producing epistemic value in the practices of data sharing in the company. The chapter provides insights into data sharing and analysis processes in the companies that are pervaded by digital security mechanisms and tactile practices of dealing with digital security. This shifts perspectives somewhat from practices enacting digital security that I have carefully distilled from everything else that was going on in the companies to a focus on the "everything else" where digital security comes up and is productive in various ways.

The previous chapters have illustrated how testing and caring perform modes of doing digital security that are rule-abiding, subverting, or a bit of both. This chapter turns to the question of how digital security is enacted as a practice of preparing, crafting once again particular ethics of digital security that differentiates what actions are considered "good" and "bad" for an anticipated future. The question that ensues is what is digital security when it is enacted as preparedness? What consequences does that have for how it can be tackled, what moralities are invited, and what relations are considered relevant?

This chapter depicts the practicalities of a digital security training. The training performs digital security as an organizational effort: an incident that is not only imagined as contained in technologies but as accompanied by organizational questions: it needs coordination, communication, reporting, and documenting. By foregrounding how digital security futures are imagined, they also become decentralized to more than technologies or individual responsibilities. In how digital security is enacted as the lurking emergency various actors contribute to it.

In previous chapters, I have explored methods for preventing digital security incidents. Chapters 2 and 3 have discussed multiple topologies of digital security and how to endure in the topology of fragility. Precarious, uncertain, and ambiguous digital security practices overlap and contradict, resulting in moral pressure of doing something "wrong". Nevertheless, the efforts necessitate some amount of hope that doing good (even though that might mean something different from what security officers say) can change the outcome of the situation. This morality and the practices tailored toward it assumed that there was still some room to do "right" after all.

Similarly, this chapter grounds its understanding of digital security in concrete practices, here, and before and during an educational table-top scenario exercise. The chapter pays particular attention to practices of anticipation, preparing, and coordinating of one singular digital security version during the training. The exercise defines what a security incident is, how to act upon it, and what a good response would look like. Thus, the exercise has an ontological dimension in which it is decided through practices of verification and purification what kind of digital security needs to be responded to.

In Chap. 2, it was argued that this oscillation of forms is characteristic of fluid forms but gains particular precarity when it turns fragile. This precarity becomes visible here as the training requires a concrete version of digital security to emerge. There can no longer be fluid forms, decisions have to be made, accountability allocated, resources prepared. Although these are clear signs of reaching stability (at least for the time of emergency response), the training is a theatrical act, a play that performs the "worst case" (this term is used during the training by one participant realizing that the attack had come from the "inside"). Fragility entails modes of action of preparing and anticipation, which render the future tangible while also phrasing present actions in relation to possible futures.

References

Adams, V., Murphy, M., & Clarke, A. E. (2009). Anticipation: Technoscience, life, affect, temporality. *Subjectivity, 28*(1), 246–265. https://doi.org/10.1057/sub.2009.18

Adey, P., & Anderson, B. (2011). Event and anticipation: UK civil contingencies and the space—Times of decision. *Environment and Planning A, 43*(12), 2878–2899. https://doi.org/10.1068/a43576

Adey, P., & Anderson, B. (2012). Anticipating emergencies: Technologies of preparedness and the matter of security. *Security Dialogue, 43*(2), 99–117. https://doi.org/10.1177/0967010612438432

Aradau, C. (2010). Security that matters: Critical infrastructure and objects of protection. *Security Dialogue, 41*(5), 491–514. https://doi.org/10.1177/0967010610382687

Baker, N. D. (2014). Role of explicit and implicit practices in the production of situated preparedness for disasters. *Natural Hazards Review, 15*(4), 5014002. https://doi.org/10.1061/(ASCE)NH.1527-6996.0000127

Baker, N. D., & Ludwig, L. G. (2018). Disaster preparedness as social control. *Critical Policy Studies, 12*(1), 24–43. https://doi.org/10.1080/19460171.2016.1214075

Beck, U. (1992). *Risk society*. Sage.

Burton, M. A. (1994). The Irrigation Management Game: A role playing exercise for training in irrigation management. *Irrigation and Drainage Systems, 7*(4), 305–318. https://doi.org/10.1007/BF00881557

Buzan, B., Wæver, O., & De Wilde, J. (1998). *Security: A new framework for analysis*. Lynne Rienner Publishers.

Caldwell, T. (2016). Making security awareness training work. *Computer Fraud and Security, 2016*(6), 8–14. https://doi.org/10.1016/S1361-3723(15)30046-4

Collier, S. J., & Lakoff, A. (2008). Distributed preparedness: The spatial logic of domestic security in the United States. *Environment and Planning D, 26*(1), 7–28. https://doi.org/10.1068/d446t

Collier, S. J., & Lakoff, A. (2015). Vital systems security: Reflexive biopolitics and the government of emergency. *Theory, Culture and Society, 32*(2), 19–51. https://doi.org/10.1177/0263276413510050

Connolly, M. (2012). Creating a campus based community emergency response team (CERT). *Community College Journal of Research and Practice, 36*(6), 448–452. https://doi.org/10.1080/10668920903018134

de Goede, M., & Randalls, S. (2009). Precaution, preemption: Arts and technologies of the actionable future. *Environment and Planning D, 27*(5), 859–878. https://doi.org/10.1068/d2608

Deville, J. (2021). Waiting on standby: The relevance of disaster preparedness. In *Ephemera: Theory and Politics in Organization*, 21(1). https://eprints.lancs.ac.uk/id/eprint/148931/

Deville, J., Guggenheim, M., & Hrdličková, Z. (2014). Concrete governmentality: Shelters and the transformations of preparedness. *The Sociological Review, 62*(1_suppl), 183–210. https://doi.org/10.1111/1467-954X.12129

Dourish, P., Grinter, R. E., La Delgado Flor, J., & Joseph, M. (2004). Security in the wild: User strategies for managing security as an everyday, practical problem. *Personal and Ubiquitous Computing, 8*(6), 391–401. https://doi.org/10.1007/s00779-004-0308-5

Emerson, R. M., Fretz, R. I., & Shaw, L. L. (2011). *Writing ethnographic fieldnotes* (2nd ed.). University of Chicago Press.

Haraway, D. (1988). Situated knowledges: The science question in feminism and the privilege of partial perspective. *Feminist Studies, 14*(3), 575–599. https://doi.org/10.2307/3178066

Haraway, D. J. (2016). *Staying with the trouble*. Duke University Press.

Heidenstrøm, N. (2019). Informal household preparedness: methodological approaches to everyday practices. *Journal of Risk Research, 23*(3), 379–397. https://doi.org/10.1080/13669877.2019.1569106

Jackson, S. J. (2014). 11 Rethinking repair. In Gillespie, T., Boczkowksi, P. & Foot, K. A. (Eds.) *Media technologies: Essays on communication, materiality, and society* (pp. 221–239). https://doi.org/10.7551/mitpress/9042.003.0015.

Kemmer, L., Kühn, A., Otto, B., & Weber, V. (2021). Standby: Organizing modes of in|activity. *Ephemera, 21*(1), 1–20.

Killcrece, G., Kossakowski, K. P., & Ruefle, R. (2003). *Organizational models for computer security incident response teams (CSIRTs)*. Carnegie-Mellon University, Software Engineering Institute. Retrieved from https://resources.sei.cmu.edu/asset_files/Handbook/2003_002_001_14099.pdf.

Knox, H. (2020). *Thinking like a climate*. Duke University Press.

Kocksch, L. (2020). Knowing IT-security emergencies. Die Herstellung von Unsicherheit. In: *Astrid Wiedmann/Katherin Wagenknecht/Philipp Goll/Andreas*

Wagenknecht, Wie forschen mit den »Science and Technology Studies«? (pp. 175–210). Transcript Verlag. https://doi.org/10.14361/9783839443798-007.

Lakoff, A. (2005). From disaster to catastrophe: The limits of preparedness. In *Online forum and essays*. Social Science Research Council.

Lakoff, A. (2008). The generic biothreat, or, how we became unprepared. *Cultural Anthropology, 23*(3), 399–428.

Law, J., & Mol, A. (2001). Situating technoscience: An inquiry into spatialities. *Environment and Planning D: Society and Space, 19*(5), 609–621. https://doi.org/10.1068/d243t.

Lindell, M. K., Arlikatti, S., & Prater, C. S. (2009). Why people do what they do to protect against earthquake risk: Perceptions of hazard adjustment attributes. *Risk Analysis, 29*(8), 1072–1088. https://doi.org/10.1111/j.1539-6924.2009.01243.x

Marres, N. (2012). *Material participation. Technology the environment and everyday publics*. Palgrave Macmillan.

McCrohan, K. F., Engel, K., & Harvey, J. W. (2010). Influence of awareness and training on cyber security. *Journal of Internet Commerce, 9*(1), 23–41. https://doi.org/10.1080/15332861.2010.487415

Mol, A. (2003). *The body multiple*. Duke University Press.

Poller, A., Kocksch, L., Türpe, S., Epp, F. A., & Kinder-Kurlanda, K. (2017, February 25). Can security become a routine? Proceedings of the 2017 ACM Conference on Computer Supported Cooperative Work and Social Computing. Presented at the CSCW '17: Computer Supported Cooperative Work and Social Computing, Portland Oregon USA. https://doi.org/10.1145/2998181.2998191

Shostack, A. (2014). *Threat modeling. Designing for security*. Wileys. https://learning.oreilly.com/library/view/-/9781118810057/?ar

Slayton, R., & Clarke, B. (2020). Trusting infrastructure: The emergence of computer security incident response, 1989–2005. *Technology and Culture, 61*(1), 173–206. https://doi.org/10.1353/tech.2020.0036

Taylor, A. R. E. (2021). Standing by for data loss: Failure, preparedness and the cloud. *Ephemera: Theory and Politics in Organization, 21*(1), 59–93.

Vinsel, L. (2023). *"How to be a better reactionary: Time and knowledge in technology regulation" Essay for Workshop "Politics of Controlling Powerful Technologies"*. Oct 17, 2023. https://sts-news.medium.com/how-to-be-a-better-reactionary-1630b5098fbc.

Williams, P. A. H. & Coles-Kemp, L. (2019). Teaching as a collaborative practice: Reframing security practitioners as navigators. In Zhigeng Pan, A. D.,

Müller, C.W., Mingmin Zhang, A. E. R., & Kifayat, K. (Eds.) *Transactions on edutainment XV, Bd. 11345*. Springer (Springer eBooks Computer Science, 11345) (pp. 108–128).

Wolf, M., & Hall, K. (2018). Cyborg preparedness: Incorporating knowing and caring bodies into emergency infrastructures. *Medical Anthropology, 37*(6), 486–498. https://doi.org/10.1080/01459740.2018.1485022

5

Performing

Instead of striving for a characterization of digital security head on—by looking at practices that allow fixing the matter for a while—this chapter explores how digital security is performative to innovation processes. It leaves questions of what digital security is behind to focus on what digital security *does*. It traces the performative effects that various securing efforts have when developing new data analytics techniques. Digital security, the chapter argues, contributes to how the companies work with data—how they build trust in data processes and generate reliable results. The chapter follows how digital security contributes to innovation practices by *situating data*—in departmental idiosyncrasies, specific moments in time and locations of data gathering.

While the last chapter presented an almost theatrical performance of digital security that allowed interlocutors to coordinate digital security ontologies for a while, this chapter takes a different path. It aims at understanding how digital security is part of "other" practices that set out to do something else than securing, such as sharing and processing corporate data. This locates digital security in the margins, or "shadows" of other tasks (Leonelli et al., 2016), echoing a marginal and interwoven understanding of digital security as an *infrastructuring* task; often invisible to formal accounts, volatile and fleeting (Star, 1999).

L. Kocksch, *Fragile Computing*, https://doi.org/10.1007/978-981-99-9807-4_5

I have introduced a questionable dichotomy throughout the previous chapters: fragile computing was described as an additional orresidual activity: unfurling in actions that are non-functional, placed "after" innovation or stand in tension with "productive" activities. This distinction does not hold up when looking at practices of data work in the companies. Fragile computing is performative to other practices regardless if its outcome is to "secure" or not.

Against the prevalent perception of digital security as hindering development practices, slowing down innovation, or holding back the free flow of data and exchange of information, the chapter argues that digital security practices are supportive or productive for *epistemic practices.* By pointing to how security practices produce *trust in numbers*, *aid in developing long-term collaborations* and *engender methods of reproducibility*, the chapter unfolds how digital security contributes to the sharing and processing of data. Fragile computing lies withing process of innovation rather than at its margins. This becomes visible when following the process of introducing a new data infrastructure in the companies engendering the innovative use of Big Data and new data analytic techniques. The companies were in the middle of developing a central data infrastructure for this purpose. These "Data Lakes" were filled with data that was formerly contained to department-specific data environments. This centralization was pivotal to Big Data efforts where old company data is given new value through processes of re-combination. By offering an infrastructural perspective on the introduction of Big Data, I aim to highlight the performativity of securing and of enduring fragility rather than jumping to conclusions on data protection or the preservation of individual's data rights (important conversations I will get back to later in this chapter).

Rather, this chapter attends to the performativity of digital security. It asks how digital security is involved in and shapes the ways in which knowledge is generated in the companies. It looks at how data are shared, withheld, aggregated, circulated, and annotated, and the different digital security mechanisms that are mobilized in the process, e.g., access control, data classification, and ownership.

I agree with Niccolo Tempini and Sabina Leonelli (2018) who state that information security has productive effects on biomedical research.

They argue that by reminding scientists of ethical obligations, forging new collaborations, and making people accountable for data sets, digital security plays a productive role in the production of knowledge. Even though they identify friction between efforts to anonymize and potential values of sharing personal information in medical research, they argue that the questions posed by digital security are important for researchers to anticipate potentially social consequences. Furthermore, by providing prompts for reflection, digital security allowed them to critically question their practices.

In full support of their conclusion that digital security can actively participate in knowledge production, I will contribute to their analysis of the perspective of corporate multi-dimensional data sharing. The companies in question were in the middle of transforming their data processing toward data science methods and platforms. This provides a reason for an approach looking into the different directions in which data are shared and processed and the respective role of digital security.

In addition, I argue that first and foremost, digital security contributes a form of localizing data in the process that urges data workers to handle data carefully with respect to their spatial and temporal attachment. In my analysis—one that is heavily influenced by STS literature—I inquire how digital security practices localize data and make people and practices sensitive to data's situatedness which in turn contributes to generating knowledge. The focus of this analysis lies in the processes of producing knowledge and not in the output that is produced because it traces the steps of data processing praxiographically rather than comparing the products. Nevertheless, epistemic products and their practices are inseparable as STS has asserted, and thus the findings of this chapter on the relation of digital security and epistemic knowledge production can give some insight into how results gain stability (Latour, 1999) and trustworthiness, particularly in new data science methods (Passi & Jackson, 2018).

A central object of STS digital data research is to problematize the popular imaginary of data mobilized in big data rhetoric and media coverage. This imaginary treats data as "objective 'givens' which are 'out there' like "water or oil" (Pinel et al., 2020, p. 179) and as holding universal epistemic values (Loukissas, 2019). In contrast, STS scholars have portrayed how data production and use require "packaging" (Leonelli et al., 2016;

Tempini & Leonelli, 2018), collaboration (Passi & Jackson, 2018), and technical infrastructures, as well as their socio-technical negotiation and enactment (Bowker, 2008; Dourish, 2017; Edwards, 2010), and that data are time-sensitive (Sørensen & Kocksch, 2021). Studies have suggested perceiving data fundamentally as a question of local "settings" (Loukissas, 2019, p. 2) and related to concrete "stuff" (Dourish, 2017).

STS scholar Geoffrey Bowker proclaims that "raw data is both an oxymoron and a bad idea", pointing out that data are never "raw" in the sense that they can never be free of traces from the means and methods used in their collection and processing (Bowker, 2008, p. 184). However, Paul Edwards argues data must be understood within concrete and local socio-technical knowledge systems or "knowledge ecologies". Knowledge ecologies are concrete, and material sets of actors equipped with particular resources and interests in a certain place and time (Edwards, 2010).

STS scholar Yanni Loukissas (2019) places emphasis on the "locality" of data. He contends it matters *where* data are produced, the specific local settings of their assembling that often differ from where data is put to use. This discourages ideas about data as independent and universal and points to questions of "where do data direct us, and who might help us understand their origins as well as their sites of potential impact" (p. 3). For data to be useful, it requires an understanding of the data's local context.

Loukissas (2019) reiterates Paul Edward's point when stating about an ecological database: "scientists, specimens, and information infrastructures are all necessary to generate, verify, and sustain what the place knows. It is the encompassing place—of which data are only a part, along with the people and plants—that holds knowledge" (p. 35). The locations in which data is generated cannot be stripped from the data, and if they are, data lose epistemic value. "When data do seem to confer transparency, it is because we are shielded from important details about the context of their creation of display" (Loukissas, 2019, p. xv). Efforts to shield the locality of data can have both epistemic and social effects as data lack important spatial detachments and can have dramatic effects on how a place is represented.

By pointing toward the context of data, Loukissas asks us to look closely at the nuanced work practices that contribute to making data seem distant and objective. He shifts attention toward the work that goes into the cleaning of data from local specificities: by packaging, standardizing standards and terminology, and deleting duplicates, data are made immutable and universal (Loukissas, 2019). Data do not come abstract, rather their local ties cause "friction" (Edwards et al., 2011).

Like Bowker, Loukissas demonstrates that the goal is not to have "less" friction by removing the locality of data to produce "raw" or "pure" results, but rather to engage actively with the concrete conditions and locations of data gathering. Against the backdrop of big data and data science, where data are considered a placeless and raw resource, Loukissas and others suggest speaking of data in plural. Data *are* locally attached, situated, and haphazard. Loukissas suggests, furthermore, to analyze "data settings" rather than "data sets" (Loukissas, 2019, p. 2) to account for concrete work that is done to put data together, and the "frictions" that emerging in the process of assembling data.

For this reason, it is necessary to observe more closely the way place is integrated, overcome, and reintroduced in corporate data practices, and particularly data science ventures. Hence, this chapter takes a closer look at data practices in data science "laboratories".[1]

Specifically, it investigates the role of digital security in the "place-making" of data in the process of subscribing data to universalist ideals, essential for the Big Data method.

In a similar, yet corporate-oriented study, CSCW scholars Samir Passi and Steve Jackson (2018) describe concrete and situated practices of data science work and unpack how data scientists produce reliable results and confront mistrust in data and data collection processes. They find data science to be a situated process of building trust in data by collaborating with domain experts and data producers. In contrast with portrayals of data science as pure number-crunching, they emphasize its collaborative, interpretative, and intuitive dimensions. Besides calculation, data science needs negotiation, e.g., what results fit with data's local origin and collection mechanism. Data, Passi and Jackson illustrate, are not trustworthy per se, but data scientists build trust in data through their data work. This includes that data science is not "number-crunching" but includes

[1] The analogy is emic. The data science teams worked in what they called "labs".

negotiation, collaboration, and careful application of data to different circumstances (p. 23).

The chapter is structured in two parts: The first illustrates how digital security shapes epistemic processes of sharing data and aggregating data at Covy's controlling unit before the new infrastructure is introduced. The second builds on the first as it discusses how this shaping was transformed with the introduction of the *Data Lake* infrastructure—a lengthy process that was not concluded by the end of the ethnographic stay. The Data Lake infrastructure challenges some of the ways in which digital security has shaped the epistemic process in the first part.

The chapter describes how data practitioners deal with specific concerns in working with data, namely, how to locate *good* data and how to build trust in data and results. Interlocutors face moments where data are indiscernible, need aggregation, and must become traceable; and situations in which data are potentially harmful, overwhelming, or lack organizational authority.

Digital Security in the *Fachbereich*

In the companies, an existing challenge was to combine data from different departments even before data science was introduced. While the departments had distinct organizational roles and tasks, their design also had epistemic consequences.

To emphasis this point, I will employ the emic term for domain, "Fachbereich", which puts an emphasis on different disciplines, or "Fächer" that lay the ground for the division of labor in the company. While "Fach" (singular) translates to discipline or domain,[2] "Fachbereich" evokes a territory, or area ("Bereich"), that this discipline is

[2] The term "Fach" is most prominently used in the school education system where it refers to a subject. At universities, the "Fachbereiche" denote the modernistic separation of independent domains of knowledge (nature, science, social science, engineering, etc.). "Fach" also has a material association; it can be a compartment or drawer.

knowledgeable in. This regional reference is worth noting when being interested in how Fachbereiche *secure.*[3]

I employ the term for a second reason: A Fachbereich in the companies is not equal to an academic disciplines or domain (such as social science, humanities, or natural sciences); they comprise interdisciplinary competencies required to attending to company functions. A Fachbereich in a German company is grown historically and their practices show a certain resistance to being affected by the latest data trends. Examples from my field would be IT or Sales, Life Insurance at Covy, or Renewables at Powino.

A Fachbereich constitutes a distinct "knowledge ecology" (Star & Ruhleder, 1994) as they differ in their choice of technologies, data practices, methods, data formats, and norms depending on and enacting their respective data objects. So, while Sales would use persona modeling based on social media data or cookies from website visitors, actuaries use statistical regression models to perform risk assessments of customers or insurance objects. At Powino, while the grid Fachbereich requires consumption data to allocate electricity needs, it must also monitor the development of electricity on the grid to ensure it does not get undercharged or overcharged, etc.

Fachbereiche (in plural) employ different kinds of data, different technologies to store data—hard drives, disks, floppy disks, basement data centers, ftp servers, or for very old data, paper—and different methods or techniques to work with data—statistics, regression analysis, monitoring, reading. More than simply supervising a different set of material devices and skills, Fachbereiche has decisive language and ways of encoding data, e.g., by naming columns with which they curated their data. For example, some Fachbereiche—especially those working with international partners—would employ a U.S. date format ("mm.dd.yyyy"), while others would use the European system ("dd.mm.yyyy"). Lastly, a Fachbereich cultivates specific ways of securing and granting access to data. This also depends on whether they had direct access to the database or were themselves users of third-party data, e.g., from international collaborators.

[3] Those interested in topologies of academia might find this regional distinction interesting too. See Chap. 2 for a discussion of digital security's reliance on regional patterns of relation.

Data securing also depends on the criticality of their data, e.g., controlling would have access to employees' banking information while Sales might have access to data on website visits or cookies: whereas the sharing of data would be almost impossible for the first, it could be possible under restrictions of the latter.

Fachbereiche are crucial because they help with the division of labor and structure accountability and responsibility, but they also serve as an organizational form that allows for heterogeneity and independent work procedures. The above-mentioned examples illustrate that data were locally bound up in the Fachbereiche, becoming accessible only through this locating within the specific organizational, technical, epistemic, and security context that the Fachbereiche provided. Data relies on its "data setting" (Loukissas, 2019, p. 2).

The term "Fachbereich" leads us to think of the unity of subject expertise, technological-organizational forms, and practices, which is useful for a better understanding of the way knowledge emerges from these specific organizational units. The existence of discernable Fachbereiche (pl.) in the companies was pivotal. Through out conversations in the companies, the Fachbereiche were used as point of reference when discussing levels of willingness to introduce data innovation or overall compliance with digital security regulations. The Fachbereiche were the primary "customers" for digital security and the data analytics teams alike. Both are considered activities across Fachbereiche (Germans have a term for that too: "fachbereichsübergreifend").

STS scholars recently discussed the purviews of non-English terms in (Law & Mol, 2020; Chernysheva et al., 2020). They argue that introducing locally specific terms may open avenues for different metaphors and writing styles. The German term "Fachbereich", I suggest, can help us to think about the unity of epistemic knowledge and specific material and spatial knowledge ecosystems better than the more common term in English-speaking publications "domain". Introducing a German term to this discussion situates the following descriptions within German companies that are characterized by a rather strict separation of Fachbereiche. This is due to the fact that they are historically grown by purchasing other companies that then often remained Fachbereiche in the new bigger company or by separating duties and responsibilities.[4]

To put it differently, the epistemic practices and expertise of the Fachbereiche are required when employing the data. Data scientists realized that they need this expertise to understand data and make useful conclusions. In turn, this means that data science solutions cannot meet a universalism but must fit with the Fachbereich-specific practices. Fachbereiche provide the socio-technical assemblages in which data unfold their full potential. So far, however, it is unclear what role digital security measures play in the process of doing data science, which is something I tackle in the second part of the chapter.

As the term "Fachbereiche" suggests, departments do not only fulfill distinct functions in the division of labor of an organization, but they have an epistemic function. Fachbereiche contribute to how companies know their business and the world they interact with. This, in turn, means that data science applications must not only fit with Fachbereich-specific work practices but must fit with their specific epistemic access to data. To make this point at the very beginning, this Fachbereich-specific knowledge is negotiated and enacted when attempting data sharing: Where can data go? What conclusions can others draw? Questions that are fundamentally shaped by digital security measures (where data can flow) and underlying data protection (where data is allowed to flow).

Digital security in the Fachbereiche was confined to two separate yet related efforts: negotiating infrastructural access and engendering epistemic access (Tempini & Leonelli, 2018). For the first, Fachbereiche employed specific hard drives or server space that could only be accessed from within, this was ensured by strict access control policies. Epistemically, only a few specialists knew how to read and make sense of data. Since data were curated within the epistemic environment of the Fachbereiche, terms in use to name columns or rows or date formats

[4] Both companies had the tendency to employ English terminology in German everyday conversations. Specifically those terms that were associated with innovative data processes or "data science" (I never heard a German translation for it) were solely in English. On the contrary, terms related to cybersecurity ("IT-Sicherheit", "Datensicherheit", etc.) as well as to the seemingly "old" organizational structure were in German. This resonates an inherent tension that was reported mainly by data scientists of incommensurable logics here demonstrated by a different semantic.

employed as well as regulatory specifications were often only known to people handling the data every day. Daniel, a controller at Covy, explained to me: *"it is not enough that you get a program, and you get access to a directory, but really that you bring some subject knowledge; what is the data and what does it [sic] do? And, what purpose is behind it?"*. Daniel refers to "subject knowledge" (in German, he uses the term "Fachwissen") that is required to understand data. Even if you gain infrastructural access to different data types, as he does in his position in Controlling, it is necessary to know how to read data and understand the local origin and relevance.

The Fachbereiche present their function as "infrastructural-epistemic units": The infrastructural access is in line with epistemic access meaning that those knowing how to read the data are also those having access. This unity of epistemic interests and data infrastructure is enabled by technical security mechanisms. Access control mechanisms or physical security controls (e.g., in the case of data stored on somebody's computer) ensure that data do not flow out uncontrolled from the Fachbereiche. Digital security thus aids in enacting the Fachbereiche by providing distinct access mechanisms. Subsequently, the Fachbereiche also enact digital security by delineating fixed corpora of data and respective access and responsibility. Digital security contributes to the separation of Fachbereiche and the infrastructural-epistemic unity, which, in turn, contributes to the organizational structure of the company. Digital security demonstrates organizational productivity.

Within infrastructural-epistemic units, it is well-established how confidential and critical data are and under what condition they could be shared. The Fachbereiche guard their data by both controlling the infrastructure and their sense-making. Digital security fulfills an organizational task and performs the separation of Fachbereiche by keeping data apart and located within them. This, in turn, has epistemic effects on the way data are composed in Fachbereich-specific methods and practices. But digital security's performativity of epistemic practices is more far-reaching than that. Not only does digital security support the unity of epistemic and infrastructural security, but it also facilitates routines to trespass this unity.

Digital Security Produces Infrastructures for Finding Research Questions and Building Collaborations

As indicated above, Fachbereiche maintained strict infrastructural boundaries. In order to transfer data between Fachbereiche before a centralized Data Lake was introduced, infrastructural and epistemic boundaries had to be overcome. Fachbereiche were separated (1) spatially—they had offices in different buildings or even different company locations; (2) infrastructurally—Fachbereiche curated their storage systems and employed a range of different statistical models and programming languages (SPSS, R, or Python); and (3) epistemically—they used different methods and strived toward diverging epistemic goals.

Sharing data between Fachbereiche was done (1) vertically and (2) horizontally. (1) Data were aggregated from the Fachbereiche to report to Covy's management. This vertical data sharing from within Fachbereiche to "higher" units like the management can be compared to aggregating scientific data and finding stable results to publish. (2) Data were, in addition, shared horizontally between the Fachbereiche, which required it to be translated from one "ecology" or "epistemic-infrastructural unit" to another. Charlotte explains to me how the latter works:

LK: So you get a data set and pass it on?

Charlotte: Yes, but rather the other way around. I need an analysis on something, so we start somewhere. Let's say we have this kind of data, but not other data. Then we ask, can we get this data from somewhere? Can we piece it together?—And this is where it turns into big data—can we combine it with other data sources and maybe come to a conclusion of something that is not there individually but in combination? This is what we steer decisions with. It is the basis on which decisions can be made. […]

LK: So you have to ask for access to specific data?

Charlotte: We get less access as the data is actively given to us. Sometimes we get access, but it is handled very restrictively. This is important because the data sometimes are really sensitive and you have to see who—not only who does some mischief with it, but you also have to understand the data. Otherwise, somebody says: I did an analysis and the data tells this and this, and the next says, but I did an analysis and the data says this and that. And so on. They draw different conclusions from the same data, they interpret them differently and get different results.

LK: How does that work if you ask for data?

Charlotte: Customer data for example.

LK: Exactly.

Charlotte: Yes.

LK: What information do you get in addition? Is that a telephone call and you say, Miss XYZ, can you give me the data?

Charlotte: Well, it is not that easy. Because sometimes we don't know what data the other department actually has. Sometimes—most of the time—there are conversations before, where we get into contact with them and say: we need this data, what do you have? How can you deliver? What is most pragmatic for you? What is most pragmatic for us? How do we get that done? We need to make sure that the data is there in time, et cetera. There are conversations beforehand and yes, when the data gets here, we get a short info: the data was updated.

LK: Is this distinction of data for data protection reasons or for cybersecurity reasons?

Charlotte: First and foremost, it's for disciplinary reasons ("fachliche Gründe") because you cannot know everything. This means every department has its sub-departments and they have their "Fach"-knowledge ("Fachwissen"). And related to that is that if you had access to everything but you wouldn't know what the data is good for, you will make wrong conclusions. So, the data is linked to the knowledge. This is how it is designed. We have some colleagues that deal with Sales, some deal with Customers, others do Controlling, Budgeting, Expenditure and so on. We have some that deal with the communication with [the headquarters] etc. […] It is a collaboration between the Fachbereiche".

In this interview vignette, Charlotte echoes what I have emphasized before: the separation of Fachbereiche for epistemic reasons. She explains that each Fachbereich is proficient with its data and relating to data from other departments requires forging relations with other Fachbereiche. It is important to find out who has data and how to localize them in the infrastructure. In addition, data are not "simply" made accessible, but they need to be understood properly. During the negotiation of infrastructural access, epistemic work is done.

First, the existence of Fachbereiche aids the search for data. She would not call everyone in the company but those that are most likely to "have" the data in question. Further, while the infrastructure allows quick access, it is also a hindrance that data are shared without any communication about why and how the data are being used. When giving access, some discussion about *"what the data is good for"* needs to take place. Negotiating access to data is accompanied by negotiating data's epistemic value and applicable methods.

The structure of isolated departments that is reflected in data holding is not set up for security reasons, but mainly to maintain a division of labor and expertise. However, this structure has unintended security benefits in the sense that it allows data to be held where there is expertise to interpret them. The infrastructural units provide an origin story for data, which needs to be known in order to use the data appropriately. I call this "data origin" in accord with Yanni Loukissas (2019). Data origin describes the concrete circumstances, research questions, and practical methods used for the collection and aggregation of data. Each Fachbereich employs different storage systems and methods as stated above. Furthermore, and relatedly, data have varying purposes. And these purposes must be understood to approach the data appropriately. The separation of the Fachbereiche—through technical measures of creating isolated data silos—hence supports the epistemic process because it locates where data comes from and fosters conversations about methods and origins. Negotiating infrastructural access performs an epistemic task in Charlotte's case. It helps her to know what the data are about and how she can use them.

Infrastructural security—the technical devices that ensure that data stay in the Fachbereiche—works as a gate between Charlotte and her colleague. Only when a conversation takes place about what data can be used for, are data shared. Besides aiding in understanding data better, infrastructural security also requires collaboration between Charlotte and her colleague which fulfills a social task. The colleague trusts Charlotte's competencies and that she will employ data in the way they are intended. This trusting relationship may strengthen collaborations in the future; it enhances Charlotte's knowledge of who has access to what kinds of data; knowledge she may later mobilize when she needs other data sets.

Unwillingly, yet nevertheless, the organizational infrastructure and infrastructural security of the separate Fachbereiche provide a means by which to locate data and access control protocols, which provoke identifying relevant questions and understanding data's origins and methods. Without isolated data storage, data would simply be accessible without prompting a discussion of the use this data can have. I understand this to be an epistemic effect as a discussion is engendered through the separate data storage as to how to use the data and what epistemic relevance they have. Data's locality is enabled through technological measures of digital security, that is, access control and separate data silos.

The first way in which digital security has an epistemic effect lies in the separation of data storage that helps to make data retrievable, understandable, and explicates origins and methods. Charlotte's story demonstrates that when data crosses the infrastructural boundary between Fachbereiche, epistemic questions follow. Infrastructural security impels her and her colleague to dive into epistemic questions, which, in turn, improves the results that Charlotte can produce from the data. Digital security fulfills epistemic productivity by producing questions of data's origins, methods, and purposes. In addition, digital security fulfills a social task by requiring the two colleagues to establish a trusting relationship and prepare future collaborations.

As we have seen, digital security serves as an infrastructural barrier in sharing data between Fachbereiche and therefore performs epistemic and social tasks. It necessitates a discussion on how data can be used which helps Charlotte choose methods and produce sound results. Digital security enhances Charlotte's understanding of the data and thus improves the integrity of knowledge produced. This has shown some of digital security's productivity in horizontal data sharing. The following provides examples from vertical data sharing.

Digital Security Produces Trust in Numbers and Accountability

Daniel, who reports business numbers to Covy's management, explains that digital security is also performing an important task in aggregating data and making reporting systems more sustainable and trustworthy: Daniel's task is to translate data vertically from existing datasets and aggregate them into management reports, e.g., on the number of contracts sold or customers who have churned to competitors. He tells me about the trajectories of data translations:

> *For example, in the life insurance area, I have about 150,000 to 280,000 datasets that are delivered each month with SAS [an analysis software]. I have a directory tree: I have my directory in there but also that of the people who deliver data to me. I have their directory too and they provide me with the data in a particular drive. For example, for one month in this form of 120,000 to 280,000 data sets and I need to take them in then. There are numbers—data fields that are numerically encoded—for example, CovyLife is the life insurance branch, it has the number 002. And I translate that into the respective clear terms, so the reader understands the table in the management report. It shouldn't say 002 there but the name "CovyLife". First, I do transformation tasks when I have 150,000, 200,000 new data sets. I do not look at them in detail, but only these blocks of data that have the granularity I want. The granularity that is appropriate for the management. This means, I translate numeric codes in text or combinations of terms and then I aggregate a second time. I must compare with the raw data from time to time to make sure that the numbers I have produced during the aggregation are still right.*
>
> *The one who passes on the data must make sure the data are correct. So, you do have to make some comparison—look at the data and for example different time steps and see if they fit in the landscape of data—does it fit in the development of the last months? This is your responsibility when passing on the data. If I pass on data with a mistake, I cannot point fingers at who delivered the data to me, but I must stop the process and go to the one who delivered it and talk to them: I have to say, the data can't be right, they aren't consistent, please check and correct them. This is the credo here: The one who delivers must take care that the data that is delivered is correct. This should not come to the receiver's attention. It happens rarely that an error transpires in the end.*

Firstly, Daniel's story illustrates what aggregating data through a trajectory of steps means: In order to be human-readable, data must be translated from numeric codes into clear text. Because the management needs to stay informed, e.g., about what branch of the company has performed in what way, the serial number for a department, e.g., "002", must be translated into "CovyLife".

Then he refers to the granularity of data he is looking for. The granularity of data is its specificity. The higher the granularity, the more specific data is, e.g., individual sales. Management reports do not require every specific detail, they require a lower granularity or specificity to identify general trends. It is Daniel's task to combine data fields or individual items into a more abstract form so management can make useful conclusions. Daniel explains that he uses data to prepare reports to the directors' board on costs, capital, or human resources. Usually, the board of directors sets the field of interest and Daniel has to provide the data. This means that he does not have to do the specification work that Charlotte did by trying to find which data set fits better and what the data "is good for". On the contrary, in his case, the data sets always come from the same people, and they have established a system of sharing data.

Third, and what is interesting when wondering about the role digital security plays here, the separated infrastructural units ("*parts of the directory*") guarantee that Daniel and his colleagues do not step on each other's toes but follow pre-decided rhythms of sharing. Passing on data from one directory to another means that one step of aggregation has been concluded and another one can begin. It needs this clear moment of opening data for the next step because otherwise work would be done redundantly, or versions could overlap. Before this point, Daniel must not enter the directory of his colleagues and just "take" the data. He has to wait for them to submit them into his directory. Each version of the data is equipped with a timestamp in their metadata, which allows Daniel to go back and compare if his predecessor has produced an aggregation that fits with his outcome. This shows an ingenious way of indicating and making use of the separated silos to make distinctions between data and allow traceable data provision.

The infrastructural security of isolated depositories allows a historical trajectory of data to be established: What step predated another step. This history allows us to locate where data were processed and what aggregation decisions were made. It makes each step in the data aggregation transparent and traceable. Data and results of this epistemic process become "backward-compatible" through the separation of data repositories which is critically enabled by both access control policies and access control convention between Daniel and his colleagues.

More than simply providing a work procedure when data can be used for the next step of aggregation, this process produces a clear line of accountability and establishes trust in numbers. Each "delivery" of data to another repository requires that the one who delivers is responsible for this aggregation step. The separate directories establish individual data workers' accountability for their data work. Daniel reflects on how passing data on from one folder to another commits him to his data: "*You always make sure that the data you pass on is coherent. And that is how even though every step is one aggregation more, means more density of the data, all levels are in reach*" (Daniel interview). Daniel aggregates all data and provides them to another folder where he and some colleagues in reporting have access. Passing on data comes with the responsibility of making sure that data are coherent. In this way, trust in data is maintained through the process of moving them from one folder to another, keeping every step of aggregation separate, enabled by access control mechanisms and infrastructural security. Passi and Jackson point out that establishing trust in numbers is by far simple but requires collaboration and establishing trust work (Passi & Jackson, 2018). This becomes clear here, as Daniel reports that the system allows going back and establishing a sound data process that can be looked up later.

Every step remains "in-reach" through this intricate system of isolated data repositories. Data's history and location can be called upon in retrospect. This ensures that if someone suspects an error, it can be traced back through the different folders and identified at which step a mistake was made. The individual folders help to keep aggregation steps in reach.

> *I have this table that has about 30, 35 rows, each row is a data point. The translation table that I can store in my directory, that is the documentation. All translation rules are defined. And then I generate from one data table and four translation tables a fifth in which I have everything. In the steps following I start aggregating [combining data points or data sets to form more abstract results]. I do not need 50,000 entries, but I cluster them, set the translated table into the level of granularity [detail of data, the more aggregated data is the less granular it is and vice versa] that is appropriate for the report to management or operational for the controller. [...] The latter has a more detailed view, more granular. They have to answer questions coming from management. I offer*

> *them a level of granularity that allows them to answer questions by "drilling down", going into a deep analysis of the data to answer the most suitable. It happens that the granularity level isn't enough and then I have to go one level deeper and dig for the data—that increases the data volume, but it does allow more detailed answers (Daniel interview).*

The data that Daniel talks about differ by stages of aggregation, data is plural (cf. Grinberg, 2016). Management does not require the level of granularity that controlling requires. But it must always be possible to "drill further down" the steps of aggregation. The steps of aggregation spread out through different stages of aggregation that are represented by the isolated folders.

The security infrastructure of isolated folders maintains a strict separation of steps, which has a threefold effect here: (1) it allows for a fixed rhythm of sharing to be set and establishes a historical trajectory of data that can be called upon and traced back for analysis purposes; (2) it makes data workers respond to their data and ensures that they do quality checks, i.e., make sure that data is coherent; and (3) by allowing to trace back steps, it fosters trust in data and provides a procedure for possible mistakes. The separation of isolated units and a convention of when data is shared makes the aggregation process transparent and data more trustworthy in the process.

This is the second contribution that digital security makes to the epistemic process of Covy: The distinction of isolated repositories in which data are submitted allows steps to be retraced in the analytic process and to "drill-down" if necessary. Hence, security increases the trust in data and reproducibility of data in question, as well as shaping rhythms of committing data.

Without mentioning digital security, Daniel demonstrates how he and his colleagues apply careful ways of sharing data; they have developed a rhythm in which to access and in which to wait for data. The evidential value of data is not an essence that Daniel and his colleagues distill but actively compose through the technical separation of work steps in which data becomes trustworthy and stable enough to base business decisions on it. It requires a sound control of each version of the aggregation process in case one has to drill down again and answer more specific questions from the management. This composition and decomposition of data are enabled through the traceability and neat separation of work steps in isolated repositories, and an (idiosyncratic) access control convention.

Daniel's way of sharing data is enabled not only by a technological separation of directories but also through the agreed convention among data workers to not meddle with each other's repositories. This combination of technical functions (isolated repositories, access control, and timestamps) and norms produces data composites that can be decomposed into clear accountabilities. The composite nature of data transpires through this example once again as data's epistemic value and their integrity not as an essence but emerging from the separate technological systems, relations between different steps of aggregation, and norms of non/accessing. This is similar to Charlotte's colleague who can adapt her security mechanisms after the informal conversation has taken place and data's integrity is composited by Charlotte and her colleague's conversation about goals, methods, and purposes.

Charlotte and Daniel demonstrate two ways in which cybersecurity engages in epistemic processes. In Charlotte's example, overcoming the infrastructural barrier engenders a collaboration between her and her colleague that clarifies and specifies the purpose of data use. This process also strengthens future collaboration between the two as Charlotte proves herself competent and trustworthy to be given data. In Daniel's example, the infrastructural separation facilitates data histories and establishes accountability among the data workers which produces trusted and reproducible data aggregates. The isolated steps aggregate data while leaving local contingencies of each aggregation step in reach. The locality of data is further explored in the next section.

Digital Security Produces Local Contingencies

Digital security contributes to epistemic processes in a third way: It reminds data workers of the locality of data. Data security requires thinking of the origins, purposes of collection, and potential local consequences of data.

> *There are special catalogues of data that fit a particular question. And sometimes someone from a small Fachbereich comes and asks us something, and it turns out to be a really complex "Gebilde" [a material figure made up of heterogeneous, hard to identify partials]. Data alone do not offer an attack vector—only when they are located somewhere, they pose a risk: In the head of an employee, for example, he can spill them. Local data is material in the sense that it offers a potential attack venture. If data was immaterial and without a place, nothing could be attacked*

Knowing about the security of data requires knowing about the locality of data. This renders data less immaterial as thought of in big data rhetoric. Rather, data have a place and digital security insists on making places specific and transparent. Security often reminded data scientists of local boundaries of data, i.e., ethical questions or existing regulatory frameworks specific to data from particular locations (e.g., from grid customers). According to the latest data protection law, the purposes of the collection have to be made transparent to customers and their consent must be given.[5] It also prohibits combining data from particularly critical parts of the company with others. At Powino, a regulation called "unbundling" prohibits the use of data from utility customers for data science purposes because concession contracts forbid any repurposing of the data. This regulation localized data's origins by stating clearly that they derive from utilities and grid customers. Besides the new data protection regulation, long-standing regulatory reasons for not bringing data together also exist.

> *The problem is the law [...] and the ethical question of can I combine different data pots and make conclusions about colleagues or customers that potentially reach far into their privacy. The challenge is that maybe we have to rely more on technical processes like anonymization or pseudonymization. [...] From a technological standpoint, it is sometimes hard to anonymize data because there is still technical reproducibility. After all, you always have a document with keys or some relational condition that has to stay raw, so you can understand the data at all.*

---

[5] On a critical reading of informed consent as a performing a logic of choice in the GDPR, see Nieves Delgado and Kocksch (2022).

Siegfried points to a relation of data locality and data security: data are always located and hence can be attacked. Vice versa, security emphasizes the locality of data to adhere to potential attacks and allocate responsibility. This responsibility then again allows conversations like that between Charlotte and her colleague where they negotiated under what conditions data can be used. The relational character of data to local origins, physical features, and processes of development becomes emphasized through digital security. Digital security produces the locality of data by noting the particular setting in which it can be harmed or attacked. In that way, digital security does not perceive data as an abstract entity but as specific, in that it is related to discreet places of collection, storing, and possible consequences. However, security can also help to make use of data that is "too local", by providing automated mechanisms for anonymization, which allow better use of data from local sources (which can stand in tension with epistemic goals as Tempini & Leonelli, 2018 demonstrate).

Digital security reminds us of and introduces ethical considerations and makes data practices responsive to, social concerns and potential data misuse. Digital security protects the epistemic processes from breaching ethical or juristic boundaries. This, however, put it into tension with sharing data freely and unregulated as suggested by data science. Digital security is often perceived as being overly restrictive and damaging potential innovative recombination of data. And some will say that it is right to do so.

The epistemic relevance of security became quite apparent for me too, when I asked about how data were collected in the companies or attempted to understand the data process. It was impossible for my interlocutors to give a clear picture. My difficulties in mapping the data landscape of the companies did not exclusively lie in the limited scope of ethnography, but in the object itself, and my interlocutors had similar difficulties to me. When asking about how data security was established, however, the infrastructure became more locally tangible. Questions like: What geographies are relevant to the data landscape? were discussed in relation to where data were stored (the US, Europe, or on-premises), especially in regard to cloud technology. This hinted toward a more material vision of data than I had anticipated. These material consequences and local contexts became clearer to me through questions about security. Data become local and plural, and gain specific, distinct histories when asking about their security. There are places where data are stored "better" than in others (on-premises rather than in the US). Security engendered a localization of data infrastructures. While when asked about the infrastructure before, data scientists would draw abstract buckets and arrows, now they would argue for material servers, transmission grids, as well as regulatory agencies, and political heavyweights (i.e., the EU regarding GDPR).

Security is usually perceived as being at odds with the sharing of data (Tempini & Leonelli, 2018, p. 664). However, as Leonelli and Tempini argue, and I demonstrated in the previous examples, security plays a critical part in the sharing of data that affects the quality of research questions, aids the allocation of clear accountability for data, and thus make results more trustworthy, as well as assisting in forging future collaborations.

When sharing data in science, Leonelli and Tempini have argued, digital security leads scientists to pose critical questions in relation to data and thereby develop their epistemic interests. We can observe this in Charlotte's example where negotiating infrastructural access entails specifying questions and research interests. In Daniel's example, digital security in the form of an infrastructure of separate data repositories improves data trust and reliability. But moreover, in the companies, security requires the asking of questions of locality and local contingencies of data that are crucial to be accountable of data and consequences of recombination. Digital security fulfills an epistemic task by producing and maintaining references of locality in data.

Unlike Leonelli and Tempini's case, however, I am not concerned with scientists sharing data among colleagues within the same epistemic sphere (a discipline), and within the same social sphere (academia), but with epistemic processes that unfold in companies. Companies are not structured by communities of practice along the lines of academic disciplines; they are structured according to Fachbereiche. Data are not shared between or within disciplines, but horizontally and vertically in the company. The Fachbereiche provide infrastructural and epistemic "silos" in which they hold their data—both for protection from external hacks, and to maintain data integrity.

Let's shift to the second part of the chapter where I discuss how digital security intervenes in the new *Data Lake* infrastructure, which suggests new dynamics of sharing data which are different from those previously observed.

What the Data Lake introduces is a notion of data that is not contained by the Fachbereiche. The unity of epistemic interest and data infrastructure is fundamentally changed through the Data Lake. I have so far demonstrated that digital security has organizational and epistemic capacities in the "old" system. The third question indicated above is: How do trends toward universalization of data and localization interplay, conflict, or collaborate? By drawing on the process of introducing the Data Lake system in both companies, I foreground the role of digital security in such new data processes. In the Data Lake, data should become abstract and accessible. The data scientists designing the lake opposed *"Individual Data Processing"* ("IDV", a term they made up to mock what they perceived as idiosyncratic and chaotic data processes in the Fachbereiche) as they deemed it too uncoordinated and potentially unsustainable, when—in the perception of data scientists—only individual data workers know where data are and how to use them.

Digital Security in the *Data Lake*

The sharing of data described by Charlotte and Daniel presents a two-directional schema as presented in Data is shared horizontally between Fachbereiche accompanying conversations and attempts to compose data with respect to their local origins. Data are shared vertically by aggregating it into different levels of abstraction between and within Fachbereiche. See Fig. 5.1. Digital security contributes to both directions differently as exemplified above, yet its main contribution is to perform infrastructural boundaries: between separated data and epistemic silos.

The Data Lake introduces a radical centralization of data which demonstrates a hitherto unknown way of sharing data multi-directionally without needing to contact data workers personally and thus detached

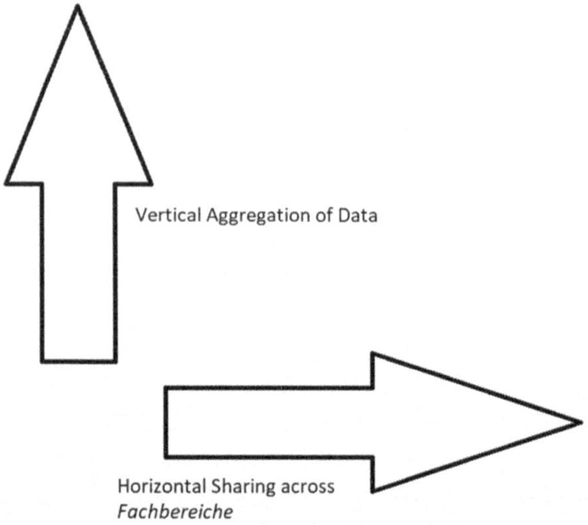

Fig. 5.1 Two-dimensional model

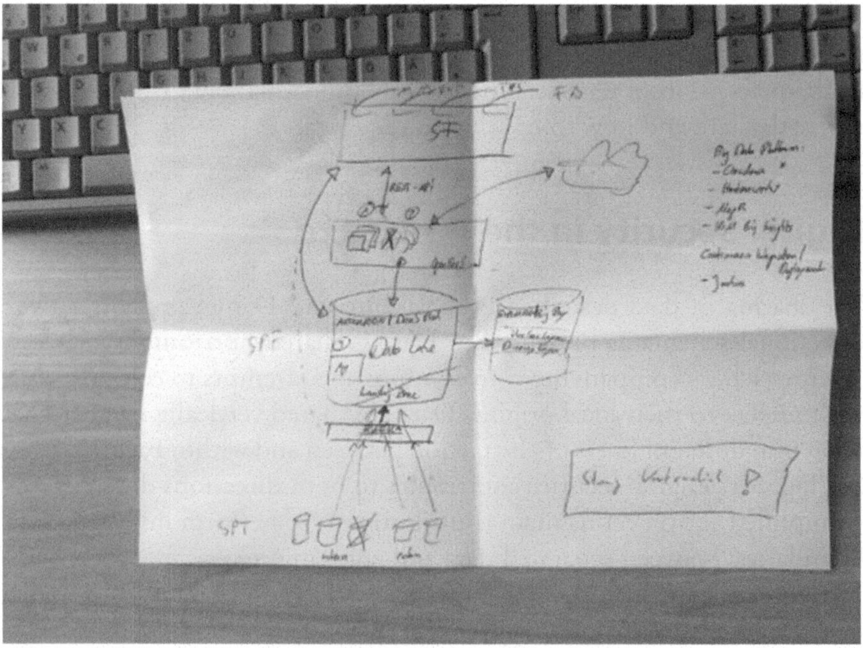

Fig. 5.2 Schematic drawing by a data scientist

from localized infrastructural-epistemic silos. Figure 5.2 present a schematic drawing by a data scientist illustrating the planned data flow in the Data Lake. Compared to infrastructures described in STS, data are not shared among one scientific community (Star & Ruhleder, 1994; Tempini & Leonelli, 2018; Edwards et al., 2009), but in a multi-directional dynamic across the Fachbereiche. Data sharing does not follow a single goal of producing robust facts about scientific inquiries but follows diverging and dynamic goals in the companies. Figure 5.3 provides a simplified version of the data scientist's graph, highlighting this multi-directional dynamic.

And consequently, the role of digital security differs significantly from scientific data sharing as the community in which data are shared and goals for which data are used are less an issue to personal conversation as they were before. The fundamental idea of the *Data Lake* was to enable more complex and multidirectional data streams within the company without having any data leak out. In addition, the *Data Lakes* provide data scientists with the easy scaling of storage, which is necessary for large-scale data endeavors, such as constant streaming from social media,

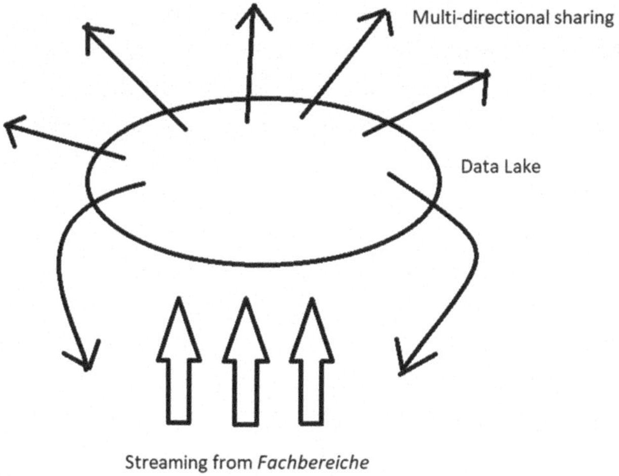

Fig. 5.3 Simplified drawing by author

or sensor technologies (smart meters, intelligent home appliances, automated grids, smartphones, weather sensors, etc.).[6]

Digital security played a crucial role in the design of the Data Lakes. It had to be ensured that data stayed within the company even when parts of the infrastructure were not self-built, e.g., when cloud technology was employed. It was also problematic to transmit data to the lakes as the moment of transmission could allow interception. When data were "on" the Data Lake infrastructure, they remained a challenge to make analysis secure, e.g., that no clear data would be transmitted in the analysis to third-party servers. Even when data were encrypted on the cloud servers, it was unclear who held the key to decrypt, potentially allowing cloud providers to have access to them.

The physical location of servers was also problematized in that regard, as servers that left the EU would fall under different protection regulations (not GDPR). Data protection created further requirements, e.g., that only some data could be processed with cloud technology, to begin with—and only under particularly digital security conditions. On the other hand, moving data to the cloud was considered more secure as data would not have to be curated on the individual—possibly not well-maintained—machines. Collecting data from small, in-office machines or hard drives and bringing it to centralized storage that ensured redundancy and backup power were considered more secure in light of damage or failure of devices, although moving data to third-party-owned infrastructures (cloud servers) caused concern. In general, reasons to move to the cloud were mainly epistemic, e.g., that new data types could be processed (streaming data from smart devices), and that existing data could be recombined. The narrative was that data's full potential could only be realized when they were combined and repurposed.

Covy struggled with the decision over what technology to employ for their *Data Lake*. At least partially data should be moved to the cloud, especially data that were generated through new smart appliances. But there was fundamental resistance against moving all of Covy's data to the

[6] Weather data was particularly interesting to both companies. Covy hoped to make better damage risk models knowing extreme weather developments and Powino planned to use climate data to project energy production by renewable sources as well as determine maintenance and repair requirements for off-shore parks, e.g., when an extreme storm or high waves occurred.

cloud, and particularly customers' data should stay within company infrastructures. A similar discussion was occurring at Powino where only some data would eventually move to the cloud. Moving to the cloud was critically acclaimed by security officers as data would leave direct—that means infrastructural—control when being stored on cloud servers. The argument was related to the technologies in use and actors having potential access to the data on cloud servers, as well as to the locality of data—as cloud servers are distributed across the globe, implicating that European data protection regulations may not apply. Both *Data Lakes* were thus designed as "hybrid" in their use of technology as both in-house servers and cloud servers were combined.

Apart from challenges in maintaining infrastructural security, the lakes also complicated the epistemic security of data, i.e., that data was still readable and could be analyzed. As I have indicated above, the separation of data in Fachbereich's "silos" had epistemic benefits and fundamentally aided the epistemic process. However, it resulted in a patchy landscape of systems, data formats, programs, infrastructures, and practices. The Data Lake was introduced to streamline idiosyncratic and Fachbereich-specific systems, tools, and practices. The silos—or patches—should come together. What Charlotte and Daniel worked with would be broken up and the data that had formerly been separated would at least partially be stored on the lake. The process of understanding datasets by reaching out to colleagues or providing a chain of responsibility as Charlotte and Daniel had done in the past would be drastically affected. Rather than discussing epistemic and evidential qualities of data with colleagues, *analytic layers* were introduced that should take over much of the analysis work. And instead of asking a colleague what data exists and how to use it, a metadata system was established. The metadata system should also contain information about the criticality of data and what should *not* be done with them, i.e., if they contained personal information or fell into a high criticality class. And instead of receiving access idiosyncratically after personal conversations or through established conventions of access as Daniel describes them, strict *access control* protocols would be established that determined who could access what data. These were set dynamically to enable rigorous control over who got access to what and for how long, which is not unlike the system introduced in Chap. 6.

The shift from a patchy landscape of silos to the centralized Data Lake was accompanied by a shift from establishing secure infrastructures toward constant efforts of securing data, as Simon, the security officer at Powino told me. As the infrastructure was no longer all "in-house", but it made dynamic use of cloud technologies and analysis applications, the data that could leave the "house" in the new infrastructure had to be put under scrutiny. This shift entailed a difference in the epistemic capacity of digital security work, shifting from determining processes and boundaries between Fachbereiche toward putting the data at the center of securing.

The new *Data Lake* introduced new data analysis procedures, and thus changed the productive role of digital security. Informal and adaptive ways of curating data access were now transforming toward a system in which access was allocated in a centralized fashion.[7] Moreover, trust in data was not reached through separately curated aggregation steps, but through parallel streams of data that could always be restored by going back to the "raw" state when needed.

In the new system, security was reached through classifying data in a standardized way, and by putting each data project under scrutiny. Although the Data Lake attempted to displace data—employing them as an abstract resource (that can stream everywhere in the Data Lake)—data science practices were not free of negotiations and local contingencies of data. But locality was composed differently, and digital security intervened in a different way. Each data set had to be classified and individual projects had to be discussed to clarify their purpose of handling a particular set of data. Instead of establishing clear boundaries between the Fachbereiche, which necessitated epistemic decisions, digital security concerns were now aimed at individual data sets. This, in turn, had epistemic consequences: less on the level of keeping Fachbereiche as epistemic-infrastructural units but on the level of providing a local context and situatedness to each dataset that was formerly provided through the infrastructural context in a Fachbereich. In that way, digital security efforts contributed to the work of data science by keeping the local ties of data

[7] But as Passi and Jackson (2018) have argued, data science does not mean solely numeric number crunching but also includes informal, collaborative, and creative practices. This is something I will not focus on here, but it was part of various occasions where data scientists negotiated and reflected upon the value of their results in informal conversations.

alive. Despite its tension with data science, digital security performed critical tasks in working with data: accounting for ethical robustness, guiding ways toward data masses, and establishing accountability. I will explain each in detail.

Digital Security Produces Robustness of Data Science Projects

The most frequent notion of digital security among data scientists was that of a "pain" or as something hemming their free exploration and recombination of data sets. This went hand in hand with a perception of old IT systems as slow and hardly adapting to changing requirements. Data scientists were sympathetic with the Fachbereiche that had to build "shadow IT" as the regular IT was too rigid and slow to fulfill their needs. The Data Lake, in contrast, could be designed more flexibly, e.g., analysis tools could easily be tested and dropped, or recombined. Georg described the Data Lake on various occasions as "tool-eclectic", meaning that the platform should be capable of offering a variety of tools depending on Fachbereich's needs.

Data scientists' work—in this very early stage of the Data Lake—was to explore what calculations would be possible with a data set. This was particularly as both Data Lake initiatives were under pressure from the management and had to prove their value. As a data scientist at Powino told me once over lunch, either in two years, everyone in the company would be employing lake technologies, or the team would be gone. To prove their worth, data scientists would employ "dummy data", e.g., openly accessible data from public offices to "play around" with analysis tools.

Dennis, a data scientist, presented one of his data explorations at a monthly meeting with data scientists and the Fachbereiche. He used data from the city government on where parking tickets were issued in the past year and analyzed them with a new tool. While data access was possible through the public office's application programming interface (API) and the publicly accessible data did not cause digital security trouble being integrated into the Data Lake, during the meeting a brief moment of confusion happened.

Dennis explored different visualization options. He visualized what type of car (make, model, and colour) received the most parking tickets in a bar chart, and in what street and at what hour most parking tickets were issued based on a simplified map graphic. During the live demonstration, the audience was euphoric, and many started thinking about possible use cases for such a tool and the data. In the middle of this euphoria, someone in the audience created a moment of awkwardness that was responded to by humor. He had realized that at least two of the tickets were issued to him.

Without further ado, the conversation continued, but later on, the fact that one person had found himself in the data was reason to consider whether the tool could be used without any ethical considerations.

While this story reverberates issues of data protection and reidentification, it still shows that security considerations were issued and did have an effect on what tools were considered for the Data Lake. Presumably, such a tool could have raised mistrust in the company if they had used it for insurance data. Such moments of awkward hesitation showed that security issues that came up during the presentation of the tool anticipated later social responses and possible consequences, which, in turn, provided a testing ground for data science experiments.

While this tool was considered a toy project that Dennis would "play around" with, other tools were not presented to Fachbereich's audience and lacked this (considerably small) scrutiny. Some of the projects that data scientists continued with were later cancelled by the data protection or digital security officers because they would breach ethical or regulatory rules.[8] Some experimental projects were withdrawn early on because data protection and digital security doubted their scalability as they violated data security. In that way, digital security and data protection concerns issued by people in the Fachbereiche and also by officers contributed to anticipating social consequences and hence built ethical robustness and longevity of data science endeavors, potentially saving them from

[8] I am not allowed to report on one project as it was first considered intellectual property of the company and later cancelled by data protection officers.

spending resources on projects that would never be realized and supporting the goal of proving their worth to the company.

Data scientist Benjamin pointed out toward the end of my stay that in retrospective over the past year projects were very exploratory, and only over time did they build the competencies to know what projects could be realized and which could not. This learning curve became apparent in the story of a digital security and data protection review board that was implemented at Covy soon after the data analytics team. Each new project had to be presented in front of the board, allowing officers to raise concerns and give advice. Benjamin explained that while discussions were rather extensive in the beginning, over time, data scientists became better at proposing less controversial projects, learning to propose more secure procedures, or at least becoming better in formulating their ideas in a way that digital security officers would agree.

While the latter might mean that data scientists have gained the competency to "hide" their security-critical ideas better, Benjamin also revealed to me in a personal conversation that while he had not thought digital security to be important at the beginning of my study, he began to understand its relevance later on.

Furthermore, during the council meetings, digital security officers pushed data scientists to formulate their projects goals clearer and make methods and goals more transparent which often "were not very clear" (Siegfried) when the meetings started. The active scrutiny by digital security officers structured the often open-ended and scarcely specified process of data science. Benjamin, in turn, told me that preparing a "case" for the council helped identify what exactly the knowledge interest was and what applications it could find in the company.

Digital security and data protection made data scientists anticipate public reception and test the ethical robustness of their projects. In a more indirect way, they also enhanced the longevity of projects with early consultations, ensuring that fewer resources were spent on projects that would never be fully realized. Furthermore, consulting with digital security officers in the council made data scientists explicate their ideas and concretize their procedure contributing to clarifying often very broad explorations of data. It should be noted that during the actual data work of data scientists, digital security did not play an important role, but regular consultation with the Fachbereiche to check upon the epistemic quality of results did, which has also been observed by Passi and Jackson (2018).

Digital Security Produces Classifications of Haphazard Data

During my stay, it was not clear how exactly the new infrastructure would look and what data would be uploaded to it. Only during my stay did it slowly turn into an object of negotiation as more concrete suggestions were made on how to assemble the technical infrastructure. One of the lines in which this stabilization took place was along the negotiation of what security levels the lake should comply with, and accordingly which data could be stored there. It became clear soon that one was bootstrapped to the other: depending on how secure the lake would be data from different criticality levels could be uploaded. But also vice versa: depending on the criticality of data, the lake would have to be more secure. This "chicken-and-egg" situation allowed the project to materialize slowly as decisions were made about which data to upload and which not. Unlike previously, the decision of what could be considered secure was not when data were within the infrastructural-epistemic unity of the *Fachbereich* but when data fulfilled abstract classification categories. Consider what the security officer at Powino tells me about the relation of security levels and criticality of data that would be turned upside down by the lake:

> *Usually you would say, ok, I have this data with that security classification and they will be processed in this system. And then we can say, ok, the security classification is this and that and for this classification, you have to take these measures. But now we have to build something where there is no data at the moment. And then you ask yourself, what data can I bring to this system? Can I fill it with highly sensitive data? Are all measures taken or not? That is why we had to do it differently in this case. We looked at the measures taken and reversed what security classification is possible to upload to it. And this is what my security assessment says towards the end: There are still some minor significant deviations from baseline security and project nature … plant as … Therefore, information security will authorize data processing within the Data*

Lake for information that falls under the following criteria level regarding availability critical, integrity serious and confidentiality medium. Ok? This means that if a Fachbereich wants to upload their data, we say, ok then do a classification of your data. If it lies below these thresholds, you can do it. But if it lies above them, we would not recommend uploading the data to what is currently being built.

The Data Lake challenges the horizontal and vertical streaming of data in which single data repertoires and pathways could be specifically built in a secure way depending on the data they carried. In centralized streaming, data are treated more universally. How to secure the infrastructure determines what data can be uploaded, while what data would be uploaded also determined how secure the infrastructure has to be built. However, this conundrum also has a product dimension. Other than the assumption of a rigid system that has fixed security measures that would determine what data could be stored, to the data scientists, the lake was less definitive. Instead, they were continuously making decisions on what security measures to include in order to include more data. These decisions were rarely deduced from actual data sets but anticipated from assumptions of what data could become useful in the future. Imagining the future security needed thus had productive effects as data scientists had to think through possible future data products.

Furthermore, digital security classifications were actively mobilized and operated as a line of stabilization for the Data Lake technology. It led to questions of what data were needed on the lake and hence how to secure the infrastructure that had to be built.

Digital security classifications contributed to the epistemic process by providing classification for the masses of data that had become available. According to geographer Rob Kitchin, big data are characterized by "being generated continuously seeking to be exhaustive and fine-grained in scope, and flexible and scalable in its production" (Kitchin, 2014, p. 27). This implies a challenge in analyzing such data. Analyzing big data must cope with "abundance, exhaustivity, and variety, timeliness and dynamism, messiness and uncertainty, high relationality, and the fact that much of what is generated has no specific question in mind or is a by-product of another activity" (Kitchin, 2014, p. 100). At Covy and

Powino, this became most apparent in the fact that "old" data were used for "new" and yet unknown purposes. This included pragmatic challenges, combining different data formats, e.g., physical paper documents and social media data, as well as epistemic challenges, knowing the contexts of data collection and what these data could be used for. The latter was similar to the problem that Charlotte encountered earlier but would not be solved by lengthy conversations with data collectors, but in a standardized fashion. The assumption in both companies was that there lies the value in an ever-growing mass of data, a value that must be explored by experimental data science.[9]

Being overwhelmed with data, both companies introduced mechanisms to classify data with the help of classification sheets that had to be filled in every time data were transferred to another service, e.g., cloud services.

The settings from which data were drawn were emphasized by security through this system of accompanying datasheets. The Fachbereiche were asked to determine the criticality of their data since only they seemed capable of knowing the potential security relevance of the data. Instead of the security infrastructure being a helpful part of making sense of data, here the epistemic processes determine who knows best about data's security requirements. Sven, a security officer at Covy, told me that only the ones working with the data can determine where and when confidentiality or integrity might be threatened. During an interview, he showed me a data assessment sheet that had to be filled in by each data worker when submitting their data into a new system, e.g., the *Data Lake* or analytic programs.

> *Sven: "In most cases, the data worker can estimate, ok, here confidentiality is violated and laws are potentially disregarded. Or if the integrity of data is violated, or the accessibility. He [sic!] can say: low, medium, serious, critical, or devastating. What each of these means you can read here. And so the data worker who knows how critical his data are or what assets he deals with sorts his data into the categories. And as soon as something above medium comes out,*

[9] Both companies employed the term "smart data" rather than big data, acknowledging in part that data masses were not automatically valuable, but data needed to be smart and be combined in a smart way.

serious, critical, or devastating, we look a bit closer and do a risk analysis. And if he cannot assess it with his own estimation, he can have a look at each point here: legal, business operations, reputation, or financial impact. There are additional information with questions for him to imagine: what data do I have and what are the legal requirements? And if he went through these questions he might have a better idea what he needs to enter above".

LK: "And who does that?"

Sven: "The business because only they know how critical their data is and if there are any legal requirements. We support them as ISO [information security officers]. We go through their assessment and see if it is plausible. Because if you want to have it easy, you just put in low and then he signs it and the manager signs it [...] And that is why we have another look at it and say, hey, you have a critical system here but you put in low in the assessment. Do you really think that if that is unavailable nothing will happen? This is a rare case but sometimes the sensibility is missing".

Sven explains that he went through the classification sheets with people in the Fachbereiche to make sure they assigned the "right" level of criticality. Although he acknowledges that the Fachbereiche know best, he is also skeptical about trusting the Fachbereiche altogether. He complains that sensitivity is missing and that he observes that people would classify data "under" the level of criticality that he sees fit. Interestingly, Karolin, a data scientist at Powino, made the exact opposite experience. Since she was in charge of bringing data from previous projects into a cloud platform, she had the previous projects fill in datasheets, which often resulted in them classifying their data at the top criticality because it meant that they would not have to share their data and take responsibility for passing it on. Karolin tells me:

Nobody in the Fachbereich wanted to sign it because they were afraid that if their name was under it—at least that's what I think. And then the responsibility was always delegated from A to B. [she quotes others]: "I am not responsible. These are not my data. I only asked for the aggregated data. The responsible person is XYZ! And then person XYZ said, yes, yes, but I don't know what you are doing with the data". And then I always tried to get them to one table, but really, they never really agreed. From about 150 sheets I got maybe 30 back and the rest went nowhere. [...] The ones with a clear admin assigned were easy

> *because that's me. I could discuss that with myself [smiling]. But the others were more complicated. For example, in Qlikview (project management dashboard) the very idea was that Fachbereiche could develop their own tiles which meant that I didn't even know what data was in there. That was a drag. And one Fachbereich was in England which means they were not as reachable to me. For example, they had developed a report with data in Qlikview to help their help-desk—they had customer data in there. That was a huge issue.*

She explains that in practice, these sheets did not work. And she agrees that if someone had asked her to sign it so somebody else could use the data in ways she doesn't know much about, she "wouldn't sign right away". Eventually, she decided that someone in management should sign the sheets if anybody, since anyway the manager has to be responsible: "If you say, ok, data owner for controlling is person A and data owner for finance is person B, it is their boss after all". She explains that outsourcing of data only works for projects in which she was responsible because she went through that trouble with different data owners and managers and many other projects failed because of it. It was still a major task for her, and she eventually decided to put a status list on her door to show her boss where each data classification was at the moment, because "every day two of three bosses were in my office to ask me for the status".

The datasheets had the effect of making individual data workers or their supervisors accountable for specific data sets, something that even though similar in the previous horizontal and vertical sharing, it became a different quality here that caused intense discussions and trouble for people like Karolin.

Furthermore, as discussed above, datasheets also fulfill an epistemic purpose: Similar to the process of horizontal data sharing above where Daniel had an established convention of when to enter each other's data repository, accountability for data is allocated through the signing of the datasheets. This is not without friction because although previously the process was clear and agreed upon, it was less clear what would happen with data here. The datasheets fulfilled the organizational purpose of replacing ad-hoc and informal sharing with a standardized and formal way of data sharing. While the sheets attributed formal accountability, they lacked epistemic security: the ones being accountable for the data sets were rarely those that knew most about how to use the data. The position of data owner was introduced to appoint people that knew most about the data and were able to assign metadata to them. In comparison to the previous process that Charlotte went through—her phone call and discussing data's origins, methods, and potential use—it was now formalized in a metadata system.

The following section demonstrates how the appointment of data owners that filled a metadata catalogue contributed to getting an overview of the various data sets and allowed them to navigate through haphazard data suddenly at reach.

Digital Security Produces Data Definitions

In the Data Lake, access to data was no longer regulated by personal contact or informal modes of data sharing as it was for Charlotte and Daniel, but by the filling in of datasheets and curating of an elaborate metadata system that served as a catalogue for the Data Lake. Data was overwhelming, and it was admitted repeatedly that one of the biggest challenges was that nobody really knew what data were stored in the siloed infrastructures of the Fachbereiche.

The role of the data owner was introduced who would be capable of providing an overview of existing data, giving relevant metadata, and assigning criticality classifications to data. The introduction of data owners thus had a three-fold effect: First, data owners should be those who worked closely with data knowing how data were collected, what methods could be employed, and what questions data could answer. Getting a network of such data owners together would then include getting an overview of the data in the company. A "business catalogue" was created that should include an overview of relevant data. Third, data owners were responsible for the data similar to those signing the sheets above.

Instead of moving data from one repository to another, all data that fell below a level of criticality should be uploaded to the lake. The lake is separated into different layers, as seen in Fig. 5.1. All data "arrive" in the landing zone of the platform. A data scientist explained to me that, in this way, the landing zone keeps track of what data are uploaded and what data are not.

For the system and curators to know what data were already accessible through the lake and what these data could be used for (similar to Charlotte's request earlier), a metadata system had to be established that catalogued the data in the *Data Lake*. Instead of reaching out to a colleague to ask whether she has access to data and what this data could be useful for, the metadata catalogue had to be consulted. How well others would understand data and know what data could be used for (both legally and epistemically), depended on the work of "data owner".

Data owners were tasked with giving meta-information on the data and permitting others to understand the data—an incremental epistemic task. Being accountable for data went hand in hand with giving epistemic guidance. Data owners would decide what classification (public, internal, confidential) their data would get and how it could be handled, e.g., shared with other Fachbereiche, shared under restrictions, or even accessible to the public. The data owner should provide sufficient metadata to use data from the lake without having to contact them to get further information as it was before for Charlotte. As Karolin explained above, the role of the data owner was all-encompassing, and taking over the responsibility did not sit well with some of the data owners. Instead of using datasheets, Covy used a metadata library to curate their data.

This was particularly urgent since the new GDPR was imminent which obliged companies to give detailed descriptions of data trajectories and data uses and announce any kind of data leakage.[10] Hence, compliance with GDPR also had an epistemic relevance: Fostering responsible data owners also meant facilitating clear definitions of data in the lake.

During a meeting in July 2017, data owners were appointed and introduced to the metadata system.

Karin, a data scientist, sits down next to her laptop which is connected to the projector in the large conference room. She opens a link to the "business catalogue system". Earlier, the system had failed to load several times because technicians were still developing it to this moment. The room of about 20 data owners

[10] https://www.gdpreu.org/compliance/fines-and-penalties/

and digital security officers follows her mouse icon. "The system is only a pilot", she explains, "we have started with a few terms and will subsequently add terms and possible relations." On the left side, she shows a row of terms. She argues that, like in a library you can define business terminology. So far you can see the organizational structure here: Contracts, Damage, Life, etc. These are categories under which you can sort your terms. When you enter a new data set, you have to identify terms used, for example for the columns or rows, and identify important relations, e.g., "this is the result of XYZ". She opens a sample page: "Contracts health insurance". When opening a page, and a thus far empty field appears and some attributes that this data set is related to. In this case: "insurance customer". I assume that this attribute was chosen because in the data set there is information on insurance customers. A data owner notices the attribute as well and argues that he would have sorted it differently, for example under "contracts". Karin replies she has noticed, and this is the exact definition of work that was needed.

Moving on through the platform, Karin emphasizes that to find data in the lake "Fachbereich-specific terms become clear and universally employed". The term "Customer" would be a good example because different Fachbereiche have different types of customers: For the health insurance sector customers are persons with full names and potentially personal data, while for other areas it can be anonymous IDs." For example, in social media. "The IT department sometimes sees employees as "customers" and might refer to them in that way in a data set. What does it mean if they use the same term for different groups of people?"

The short episode from one of the several hour-long meetings on this issue illustrates how the business catalogue operates as a navigation device for the *Data Lake*. The problem discussed here is closely related to what Daniel has explained above on different ways of collecting and entering data into databases in the Fachbereiche. His problem of combining "apples and oranges" comes up here as one term is used for different things in the new system. Karin and other data scientists explained to me that they would have to develop a "dictionary" along with the *Data Lake* to enable a universal understanding of data in the lake together with Fachbereich experts.

Consequently, the data owner did not only the role of being accountable for data (through a signature on sheets) but also that of a "curator" for the data who knew them well enough to be familiar with their local specificities. Karin explained that data owners needed to know "where data is around in the daily business" emphasizing their locality as a description of what practices data are part of and what local specifics need to be considered.

The data owner system thus assisted in localizing data or perceiving the respective "data settings" (Loukissas, 2019, p. 2) in which they are composed and relevant. Different column names are the product of the concrete relations of data and must therefore not simply be eradicated but carefully negotiated. Digital security's contribution in this process was that the data owner—that was formally introduced as responsible for data—also served as a local curator.

Karin adds that this is not only about terms but also more generally what information is in the data and how to find them. And what regulations data may fall under. She adds that regulations, for example, data protection regulations or digital security regulations could be marked as relevant here: "You can write something like 'please comply with regulation X' or 'not GDPR relevant'". "The system is a way marker to the lake", one data owner realizes. Karin acknowledges that sometimes the whole data set is GDPR not relevant, but that it was usually considered "contaminated" if only one field or line was critical. "It has to be an open discussion how to handle this here, particularly with 'crown jewel' data". She adds that the problem can not only be located in one single data set but emerges from the combination that becomes possible now. A security officer adds that for example, the life insurance of politicians could cause bigger monetary damage than other insurance contracts. The most important aspects to define crown jewels are not only the amount of data but also the context of usage and the way data is lost or damaged. If a prominent politician's contracts would be lost due to negligence, this could result in higher damage than if they were hacked (a security officer exemplified). Every Fachbereich has different crown-jewel data, it is not always a prominent customer but can also be data that is critical for business processes.

The catalogue not only comprises information to understand the specific settings of data but also data protection and digital security regulations. Again, in comparison to Charlotte's practice where her colleague gave her access after a conversation and warned of potential risks of dealing with data, the metadata system operates as a place to systematically curate such information. Notably, as it was for Charlotte, getting access to data means familiarizing oneself with data's "setting" (Loukissas 2019) and their data protection and security alike.

The metadata system nevertheless "cleans" the data by curating both locally relevant terms in the system (categorized after the Fachbereiche) and instructs the creation of a universal vocabulary. The data can remain somewhat locally detached, while the metadata system must be universally understandable. While data can remain complex and unstructured (non-numeric data), the metadata system allows universal empirical access.

The unity of infrastructural and epistemic security changes in the Data Lake. In the Data Lake, epistemic access is guaranteed through the appointment of data owners who are tasked with the creation of a reliable metadata catalogue that translates the relevant contextual and local traces about data. Part of this process is the development of data criticality evaluations through evaluation sheets. Those sheets however were problematic in that they materialized responsibility for data which did not work well with data owners. While the sheets and metadata system had epistemic effects: they required data owners to get to know their data concerning confidentiality and criticality, they did not fit with the general complexity of data in the companies. It was not within individual responsibility to sort out all data into categories. Both the sheets and the position of data owners were critically acclaimed by those having to take over new responsibilities. In relation to the data owners, it was unclear whether the right people had been picked, as many argued they did not know the data well enough. Although troublesome and causing conflict, the sheets and data owners did support the process of data science by posing relevant questions and reminding data scientists of the local origins and consequences of data, which can be described as "place-making" within the data science practices of detaching data from local origins. Data owners were established based on digital security definitions and soon took over more epistemic tasks, e.g., assigning metadata. Knowing what data was particularly valuable was not a security decision alone but an epistemic decision.

Implications 8 and 9

This chapter has pointed toward an understanding of digital security not as a mere remedy to “disturbance”—failure, hacking attempts, leaks, or breaks—but as an integral part of information infrastructures and epistemic practices of sharing and employing data for innovative purposes.

Implication 8: Digital Security is Productive in Data Innovation Processes

This chapter contributes to the bigger scope of this book as it takes insights from STS—on the productivity of information infrastructures for the creation of knowledge—and mobilizes it for the study of digital security. The chapter illustrates how digital security becomes productive in the processing of company data. It argues that digital security has epistemic consequences. It aids in knowing data because it supports the sociotechnical apparatuses through which knowledge is produced (infrastructural). Furthermore, digital security operates as a devil’s advocate in order to make infrastructures more resilient and anticipate social consequences (accountability). Finally, digital security engenders critical questioning and facilitates a conversation about the local setting of data, which helps in understanding data as an epistemic device that carries forward traces of local contingency (locality). The three dimensions—infrastructural support, allocating accountability, and reminding of and insisting on the locality of data—contribute to epistemic practices and the innovation process.

Bluntly put, my suggestion in this chapter is that digital security is producing epistemic knowledge because it “slows down” (Stengers, 2017), irritates, and is uncomfortable in prompting important questions during innovation processes in corporations. I have pointed out in another chapter that digital security relates to the development and exploratory data analysis like a chicken to an egg or like a cat to a mouse. This chapter argues that, in particular, the confronting and dissenting position has generative effects that have yet to be explored further. The chapter argues that the fragility and uncertainty that is part of digital security is not to be understood as a problem—and something to eradicate—but as a value in itself. Digital security does not "only" protect values from attackers, it is itself valuable and productive. Such a productive notion of digital security is important as it points to the variety of practices and relations that emerge from fragility, practices that include productivity, in this case, of epistemic value. This contends that digital security is not solely a "hemming" or invaluable practice of maintenance, but rather that it actively produces and advances things, e.g., epistemic processes. Table 5.1 provides an overview of the ways in which digital

Table 5.1 How digital security contributes to data innovation

Infrastructural Form	Epistemic Units	Epistemic Practice	Details
In silos	Fachbereiche	Finding research questions	Urging discussion of data's origin, methods, and purposes; forging trusted collaborations for the future
		Producing trust in numbers and accountability of results	Establishing isolated directories and a convention of sharing
		Producing local contingencies of data	Showing data's local consequences and origins
In the lake	Analytic layers	Producing robustness	Asking what data science projects are compliant with data protection and what social consequences are possible
		Producing data classification	Offering a way of classifying data along with existing criteria which allocates accountability for data and helps establish trust in data.
		Producing data ownership	Overcoming overwhelming data masses, creating a catalogue of epistemic value and criticality in tandem.

security contributes to epistemic processes and data innovation in the companies.

Tempini (2016) argues that "Information Security requirements are situated and localized in each project and defined as much by the nature of the data types and phenomena in question", highlighting that information security depends on the specific setting of data (Loukissas, 2019). I have argued here that the case also works in reverse: information security contributes to localizing each project and helps to define data types and the phenomena in question; digital security produces the setting of data. As Tempini and Leonelli (2018) state on sharing biomedical data, concerns about the privacy of patients and ethical principles must be balanced with a will to publish for the public good.

Digital security affects and enables epistemic processes and shapes the way data are used to make knowledge claims. Digital security urges data workers to deal with the dependencies of data aggregation and allows traceability and data histories to be in reach, while also forging relations between data and data owners and between colleagues in the Fachbereiche. Beyond the traditional critique that digital security is an external factor to productive actions of building infrastructures and working with data, this chapter argues for a productive perception of digital security. While it might take time and trouble and partially complicate processes, it serves to make data work more accountable, cooperative, and long-term (Tempini & Leonelli, 2018). Digital security makes the new infrastructure more sustainable and resilient to fast-changing requirements that are explicitly hoped for by data scientists. Digital security contributes to turning the new infrastructure into a durable solution that works both for data scientists and their need for an abstract data pool and the Fachbereiche who still require localized and specific data.

Implication 9: Shifting Attention to Digital Security in Data Science Research

Unlike Tempini and Leonelli (2018), my analysis looks at corporate data practices that are more heavily influenced by neoliberal agendas of free fluxes of data and unlimited recombination and marketization, but that also carry enormous value for the continued work of the infrastructures they maintain. The role that digital security plays here is particularly crucial as the fears of extractivist data practices are more imminent in corporations than in science. My approach thus displaces Tempini and Leonelli's findings to corporate data practices and more specifically data science ventures that have hitherto not been described in their relation to digital security practices.

Furthermore, by exploring how security's performativity transforms through data science and new ways of organizing data (from unidirectional to multidirectional, see Figs. 5.1 and 5.2), we learn about the way epistemic practices change when data is delocalized and how digital security plays a role here. The second part of the chapter is concerned with the introduction of a platform whose design is aligned with the big data rhetoric of data. The Data Lake infrastructure introduces a centralized data storage that will replace local and contingent data systems. In the Data Lake, digital security contributes to making order in an overwhelming amount of data, while also fostering accountability with the introduction of datasheets and data owners. Specifically, the role of data owners was aimed at establishing security, but they soon turned into epistemic curators for data. While the Data Lake fosters a detachment of data from their local origins, the data owners operate as "local anchors". It must be noted that the distribution of responsibility to data owners was not without friction.

The *Data Lake* infrastructure is associated with big data narratives, making data freely flowing and universally usable. Digital security in a way disturbs that ideal by insisting on and performing the locality of data. Data science and digital security often stand in tension and appear mutually exclusive, with projects seeming as if they cannot be secure and open and innovative at the same time. In this chapter, however, I have argued otherwise. I have put forward a more dialectic and procedural understanding of the relations between data science and digital security, in which the latter contributes actively to the former. This does not deny that both are in tension, but rather than perceiving this tension as negative and hemming, I perceive it as productive and dialectic.

Data science has received considerable attention in CSCW and STS. A shift in the modes of knowledge production has been observed, reaching as far as to an "end of theory" as proclaimed by some (Anderson, 2008). Data science is being employed in a range of areas of daily life, from cost-calculation in grocery stores to policing and science (Kitchin, 2014). In this chapter, I have highlighted some of the practical consequences of data science, namely the need to develop new data infrastructures (the Data Lakes). The chapter has argued that data science is not solely a shift in techniques, but requires fundamental changes in how data is stored, shared, understood, and secured. And more precisely, the chapter has argued that digital security plays a crucial role in the building of new data infrastructures and in producing knowledge from them.

Through digital security, data scientists build accountability in data and facilitated epistemic insights. This finding contributes both to the existing literature on data science practices and questions of how to establish sound results (Passi & Jackson, 2018), as well as to literature that is interested in the ways data are localized in epistemic practices (Loukissas, 2019). Digital security has only very rarely been considered part and parcel of epistemic practices (Tempini & Leonelli, 2018), and this chapter has aimed at this gap by describing the localization work that digital security does in epistemic processes, particularly during a time of transformation. Digital security provides a means by which to slow down and critically question data science, making it a relevant tool for critical analysis of the data science promises.

In short, digital security introduces "friction" in data science. This "data friction" reminds data scientists of the local and historic origin of data (Edwards, 2010, p. xvii). As I have shown here this can have productive effects for data processing. The difference between ways in which digital security produces knowledge in siloed infrastructures and centralized infrastructures have been outlined above. They lie, specifically, in the configuration of expertise and the forming of a universal catalogue. Passi and Jackson (2018) show that both too identical and too different results to old methods are scrutinized in data science ventures. Domains are reluctant to adapt new analysis tools if outcomes are too similar to previous techniques (as they see no value in changing routines) and if outcomes are too different (as they seem unreliable) (p. 19). The correct amount of friction through data science is needed. In my examples above, digital security aided this calibration work, making processes and vocabularies both similar and specific enough. It is important to look closely how digital security configurates the additive value of data science ventures, and where and how the configuration of access causes productive friction.

Authors have argued that current research on data science has left a blind spot for existing corporate data science work (Knox & Nafus, 2018). Due to a lack of access for researchers, corporate data science practice is only very scarcely described. And if so, research is limited to small-scale and start-up applications (Passi & Jackson, 2018). During my ethnographic research, I was given the opportunity to observe the introduction of data science in two large-scale companies. This insight into corporate practices advances our understanding of the infrastructures of data science—both their material labs and arrangements as well as the computational infrastructures that equip data science. This chapter has introduced the Data Lake infrastructure as an exemplary case in the two companies. Data science in large-scale companies requires a fundamental reorganizing of data and data ecosystems (the Fachbereiche). I have portrayed it here as an infrastructural effort, encompassing new data, tools, organizational roles (data owners), processes, and more. While digital security and data science are often conflicting, e.g., on the dynamics of employing devices and data, digital security contributes and is performative to the infrastructural work that is part of data science ventures, e.g., by slowing down and localizing data, thus building accountability, and classifications necessary to handle new data types and consolidated data sources.

Finally, this chapter offers insights into the role of digital security in various stages of "infrastructuring" (Star & Ruhleder, 1994; Pipek & Wulf, 2009) data science, from the design and sketching out to the implementation and maintenance of a new information infrastructure. This contributes the dimensions of digital security and data protection to the existing literature in infrastructure studies. The introduction of new data infrastructures has long been a topic for STS. Leigh Star and Karin Ruhleder (1994) followed the introduction of a new data repository (the "Worm Community System") among genetic biologists and found that it had to overcome various socio-technical challenges. For example, the technological format in use (UNIX) was not familiar to scientists and they felt overwhelmed with learning a new system. Furthermore, allowing others access to their own research data was particularly a problem to younger scholars who still needed to publish the data for their qualification work. Star and Ruhleder find that the introduction of infrastructures comes in the form of an entire "ecology" of components that must comply, thereby including both social and technological dimensions. Digital security has not yet been a focus of attention in infrastructure studies, a gap I attended to in this chapter by pointing to how digital security relates to the introduction of such a new ecology. Digital security must be taken into account when researching how infrastructures are built and maintained, this chapter has made a first attempt to do so.

Even in those aspects where digital security challenged the epistemic process, e.g., by prohibiting data combination, it ultimately stabilized the data infrastructure by making it more resilient against adversarial attacks as well as public skepticism. While I have shown the infrastructural and epistemic effects of digital security, it can also contribute to justifying or "greenwashing" (or "secure-washing") data science endeavors. While this stimulates a longer debate about the ethical dimensions of digital security, it is a noteworthy finding that digital security by acting as an agent of friction in the process of data science aided the stabilization and resilience of the data science project. Digital security emphasized the locality of data and concrete data processing, thus suggesting data as situated and patchy rather than as universal resource.

While it has been established that digital security takes an active role in scientific knowledge production (Tempini & Leonelli, 2018), I have shown that this is also the case for corporate knowledge production. Hence, I have made the point that corporations should not only see digital security as a necessary side task but as fundamentally part of making knowledge claims and building robust processes. It is worth understanding how companies use data and create trusted findings that steer their decision-making. I have focused on the role of digital security in this process. Corporate knowledge practices at Powino and Covy are shaped by rhythms of access and non-access enabled by digital security mechanisms.

Companies' data are considered an "asset" that is only accessible from within the company. Calls for company transparency and responsible corporate action must negotiate with the company's digital security requirements. It is thus interesting also from the perspective of critical security studies where company interests and digital security interests overlap and what alternatives are possible. I have argued in Chap. 8 that existing exercises follow an asset-centric logic, while other practices in the companies suggest other registers (Chaps. 7 and 8). This chapter contributes to widening the logics or models in which we think about digital security, here shifting from restrictive and hemming toward a productive notion of digital security.

I have taken yet another approach of determining the practicalities of digital security in this chapter, by asking not how digital security emerges from practices, but how it enacts ways of sharing and producing facts from digital data. Fragility is particularly put into a more generative tone in this chapter, as it is not a symptom of failure but something that engenders, localizes, concretizes, questions, or produces accountability. Digital security plays a role in stabilizing facts from digital data and performs the task of a "place-maker" by emphasizing local origins and evoking negotiations of specific circumstances and consequences of data. Fragility—of potential data leakage, imperfect infrastructures, and sharing mechanisms—does not become a state to overcome but to be taken seriously and work with during epistemological processes.

All chapters have approached digital security as a shifting topology, negotiation, and part and parcel of knowing data. This is quite unusual when talking of digital security because it is often framed in terms of isolated functions, mechanisms, or incidents, or follows rhetoric of attack and defense, equipped with moral imperatives. I have refrained from narratives for the sake of painting a wider picture of distributed digital security efforts in the two companies, closely tied in with other tasks, e.g., developing a new data technology. I did not frame digital security as the avoidance of a breakdown, but more broadly as a system of practices coping with but also engaging with the possibility of breakdown. This is a small but crucial shift in perspectives, as I will explain in the following chapter. I have suggested several examples of concrete acts of *doing* digital security: testing, tinkering, training, all of which emerged only by halting and staying with practices for a while rather than jumping to explanations or judgments. It is time to put the wondering to rest now by pointing to some of the conclusions that can be drawn from the material and proposing a definition to think with about digital security in the future.

In each chapter, I have pointed to two or more implications, nine implications in total. I have furthermore employed the term digital security eclectically without pointing to distinct features or mechanisms. I do so intentionally, in order to not limit associations to technologically defined questions or specific human-machine configurations. This is a slippery slope because, on the one hand, such an approach does not predefine digital security but rather lets concrete engagements define when and where securing occurs. On the other hand, an approach like this must decide what qualifies as relevant moments of observation. For example, looking at those moments where digital security is vague, unfinished, and ambiguous. In practice, this means staying open and attentive to noticing when securing is negotiated or appears as an issue of concern. This means not approaching digital security from pre-defined settings where "it" matters according to abstract definitions, but letting practices of tackling, negotiating, subverting, and stabilizing emerge and drive the object of observation.

Much has been asked of the reader, for example, to endure sometimes long empirical vignettes and quotes. The purpose of these was to offer curated stories of digital security that would allow juxtaposing and wondering. Your thoughts while reading those might have been different from what I concluded. This is rightfully so. Other interpretations would have been possible, with other theoretical devices attached to them. My goal was to develop "theoretical technologies", as Estrid Sørensen (2009, cf. Dewey, 1994) calls it; and thereby "open up" the ontology of digital security (Liebetrau & Christensen, 2021). Theoretical technologies do not precede the empirical—they are not simply drawn from literature—but neither do they appear ominously from nothing. Rather concepts are taken to the field, they go to "work", challenging and allowing specific conclusions about the field. Theoretical technologies emerge from being in the field and studying literature in tandem, juxtaposing both in a constant process. Rather than a law or the result of a vague thought process, theories operate as tools, or "devices" (Winthereik, 2020) that are "trialed and tested" (De Laet et al., 2021) with each new field. Theorizing in STS is perceived as a process (hence the gerund form) that takes place within the telling of stories from the field.

References

Anderson, C. (2008). The end of theory: The data deluge makes the scientific method obsolete. *Wired Magazine, 16*(7), 16–07.

Bowker, G. C. (2008). *Memory practices in the sciences*. MIT Press.

Chernysheva, L., Dányi, E., de la Cadena, M., Farquhar, J., Joks, S., Kondo, S., Lai, L., Law, J., Lin, W., Magelhães, B., Martínez Medina, S., M'charek, A., Mol, A., Sezneva, O., Swanson, H., & Østmo, L. (2020). Translating a title: On other terms. *The Sociological Review, 68*(2), 283–286. https://doi.org/10.1177/0038026120905470

De Laet, M., Driessen, A., & Vogel, E. (2021). Thinking with attachments: Appreciating a generative analytic. *Social Studies of Science, 51*(6), 799–819. https://doi.org/10.1177/03063127211048804

Dewey, J. (1994). *Experience and nature*. 9th printing. Open Court (The Paul Carus lectures series, 1).

Dourish, P. (2017). *The stuff of bits. An essay on the materialities of information*. The MIT Press. https://ieeexplore.ieee.org/book/7932719.

Edwards, P. N. (2010). *A vast machine: Computer models, climate data, and the politics of global warming*. MIT Press.

Edwards, P. N., Bowker, G. C., Jackson, S. J., & Williams, R. (2009). Introduction: An Agenda for infrastructure studies. *Journal of the Association for Information Systems, 10*(5), 364–374. https://doi.org/10.17705/1jais.00200

Edwards, P., Mayernik, M., Batcheller, A., Bowker, G., & Borgman, C. (2011). Science friction: Data, metadata, and collaboration. *Social Studies of Science, 41*(5), 667–690. https://doi.org/10.1177/0306312711413314. Retrieved from https://escholarship.org/uc/item/2tj4g7sc

Grinberg, Y. (2016). *Is data singular or plural?* CASTAC. http://blog.castac.org/2016/10/is-data/.

Kitchin, R. (2014). *The data revolution: Big data, open data, data infrastructures and their consequences*. Sage. https://doi.org/10.4135/9781473909472.

Knox, H., & Nafus, D. (Eds.). (2018). *Ethnography for a data-saturated world*. Manchester University Press.

Latour, B. (1999). *Pandora's hope. Essays on the reality of science studies*. Harvard University Press.

Law, J., & Mol, A. (2020). Words to think with: An introduction. *The Sociological Review, 68*(2), 263–282. https://doi.org/10.1177/0038026120905452

Leonelli, S., Rappert, B., & Davies, G. (2016). Data shadows. *In Science, Technology, and Human Values, 42*(2), 191–202. https://doi.org/10.1177/0162243916687039

Liebetrau, T., & Christensen, K. K. (2021). The ontological politics of cyber security: Emerging agencies, actors, sites and spaces. *European Journal of International Security, 6*(1), 25–43. https://doi.org/10.1017/eis.2020.10

Loukissas, Y. A. (2019). *All data are local.* MIT Press.

Nieves Delgado, A., & Kocksch, L. (2022). Privacy in social media. *Techné: Research in Philosophy and Technology, 26*(2), 257–284.

Passi, S., & Jackson, S. J. (2018). Trust in data science: Collaboration, translation, and accountability in corporate data science projects. In *Proceedings of the ACM on human-computer interaction*, 2(CSCW), 136. https://doi.org/10.1145/3274405.

Pinel, C., Prainsack, B., & McKevitt, C. (2020). Caring for data. Value creation in a data-intensive research laboratory. *Social Studies of Science, 50*(2), 175–197. https://doi.org/10.1177/0306312720906567

Pipek, V., & Wulf, V. (2009). Infrastructuring: Toward an integrated perspective on the design and use of information technology. *Journal of the Association for Information Systems, 10*(5), 447–473. https://doi.org/10.17705/1jais.00195

Sørensen, E. (2009). *The materiality of learning: Technology and knowledge in educational practice (Learning in doing: Social, cognitive and computational perspectives).* Cambridge University Press. https://doi.org/10.1017/CBO9780511576362

Sørensen, E., & Kocksch, L. (2021). Data durabilities: Towards conceptualizations of scientific long-term data storage. *Engaging Science, Technology, and Society, 7*(1), 12–21. https://doi.org/10.17351/ests2021.777

Star, S. L. (1999). The ethnography of infrastructure. *American Behavioral Scientist, 43*(3), 377–391. https://doi.org/10.1177/00027649921955326

Star, S. L., & Ruhleder, K. (1994). Steps towards an ecology of infrastructure. In J. B. Smith, F. D. Smith, & T. W. Malone (Eds.), *Proceedings of the 1994 ACM conference on computer supported cooperative work* (pp. 253–264). ACM.

Stengers, I. (2017). *Another science is possible* (S. Muecke, Trans.). Polity Press.

Tempini, N. (2016). *Science through the "Golden Security Triangle"*. Information Security and Data Journeys in Data-intensive Biomedicine: Association for Information Systems (AIS). https://ore.exeter.ac.uk/repository/bitstream/10871/24795/1/InfoSec%26Research_ICIS_2016_Final_NT.pdf.

Tempini, N., & Leonelli, S. (2018). Concealment and discovery: The role of information security in biomedical data re-use. *Social Studies of Science, 48*(5), 663–690. https://doi.org/10.1177/0306312718804875

Winthereik, B. R. (2020). Is ANT's radical empiricism ethnographic? In A. Blok, I. Farias, & C. Roberts (Eds.), *The Routledge companion to actor-network theory* (pp. 24–33). Routledge.

6

Fragile Computing

Living in a world that is damaged by loss of biodiversity, harm to oceans and lands brings the realization that we may not be able to innovate ourselves out (Tsing et al., 2017). This book suggests adopting this current line of thinking about forms of living beyond progress for matters of digital security.

If we go out into a world that is broken, what we hope for changes. We hope less for repairing or reinstating security, but rather brace ourselves to find cracks, imperfections, and dilemmas. When including digital security in this embracing of broken worlds, we are prone to fall into the fallacy of naïve appreciation for minute and small actions of maintenance. The stakes are high; so while fragile computing includes a modest understanding of what digital security can be, it is not naïve to the challenges and dangers. Fragile computing involves halted judgment, but it is oblivious to the potential disasters caused by insecure technologies. Enduring fragility sits at this awkward in-between: appreciating mundane tactics and seeing them in tension with harsh global digital security issues.

Besides stories of growth, solutionism, and purification, we find appreciation in "forging new relations" (Clark & Haraway, 2018, 5) and "staying with the trouble" (Haraway, 2016). New relations include uneasy

L. Kocksch, *Fragile Computing*, https://doi.org/10.1007/978-981-99-9807-4_6

alliances; not judging innovators for their innovations. Staying with the trouble means staying attentive to the need to secure while showing appreciation for situated mitigation. Grant solutions will continue to be designed, they will gain traction and leave traces. Living beyond progress also means living with those that continue to innovate and fix. Fragile computing is about humble mutual interrogation. It is about noticing what happens in the shadows while being attentive to needs to attune and apply. The last chapter has attested to this ambiguity by asking how fragility contributes to innovation, while Chaps. 2 and 3 have shown that fragility emerges in the disconcertment that is involved in testing and tinkering with imperfect and ambiguous security.

When thinking about digital security as fragile computing, we accept the need for such an alternative thinking. This includes thinking about maintenance rather than repair (Denis & Pontille, 2011, 2013), moral ambiguity rather than fast judgment (Mol, 2002; Stengers, 2017) and uneasiness rather than detachment (De Laet et al., 2021). We can begin to hold precious what happens in the friction between solutionist narratives and situated material practices.

Fragile computing means thinking differently about the dangers of digital insecurity; a thinking that is not aimed at solving but at enduring the various stakes at play. The chapters in this book have opened up practices to make visible this uneasy work of living with insecurity. A halted judgment is not non-judgment. It is this uneasiness where fragility can help us conceptually. Fragile computing emerges as a mode of inquiry rather than as a clearly or prior demarcated concept. It raises attention to the mundane while not being naïve to the disruptive potential and dangers emerging from digital in/security. Fragile computing dwells in ambiguity: to conflicting goods, moral tensions, and active inactivity. The book tells a story about digital security that subverts dominant logics of renewal and urgency in digital security by turning to the "pain-points" of digital technologies (Anand et al., 2018; Jackson, 2014). While painful, thinking with fragility is not hopeless. Hope lies in the ordinary, in the performativity of fragility rather than in grant solutions. It means hope in the ordinary attempts of securing, making things "slightly better or worse" (Danyi, 2022). Maybe even security experts can find some consolidation in the stories of this book it; admiring unanticipated strategies of living with insecurity that could be strengthened. Others may end up frowning. In fact, these ambivalent reactions are what are worth exploring further. It is an uneasy and perpetual task put upon digital security research to include fragility. It cannot mean "just" including ethnographic methods in the repertoires of digital security research but including other modes of thinking and being in the field. Doing this work shall leave researchers disconcerted, not with ready-made judgment in hand. It may mean letting things perish and relating to what perishes, it will be uneasy to think about digital security as fragile.

While sensitizing us empirically, fragile computing also opens a world of digital security practices so far underestimated, or, even more importantly, deemed incompetent or unaware. The book suggests widening what is perceived of doing digital security to practices of engaging with fragility. Such an "ontological opening" (Liebetrau & Christensen, 2021) provokes and troubles. The terms associated with this troubling have been care, preparedness, and performativity. They were meant to "[u]nravel tensions" (Mol, 2006, 412) that lie within fragile computing. This opening is ongoing. Accordingly, this last chapter will not provide closure but rather guide and consolidate where a consecutive debate might be heading.

I have so far been hesitant to provide definitions or follow definitions of digital security that are common in the field. I have done so to leave space for wondering and investigation of the multiple forms and practices involved with fragile computing. This hesitance does not imply that we should avoid definitions altogether. On the contrary, definitions provide ground for further specification and discussion. While it was important to remain unspecific to explore unanticipated and unexpected dimensions of digital security that may have otherwise been neglected or assumed unconnected. Now it is time to be more specific. We see many relations are relevant to digital security that we (including me) have not seen coming. What results from fragile computing? What are the things to "notice differently"? (Tsing, 2015). Where to "slow down" our reasoning on digital security? (Stengers, 2017).

Fragility lies beyond narratives of total breakdown and smooth progress. Denis and Pontille write: "consider the fragility of the world beyond the figures of disaster and collapse" (2020, 15, translation by author) to point to the normality of fragility that especially maintenance workers are accustomed to. Ethnographic attention to failure and breakdown has been prominent in recent anthropology; beginning with Tsing's (2004) attention to "frictions" in global markets to Appadurai and Alexanders' (2020) proposal of failing systems and imaginaries.

Authors have acknowledged the plethora of practices that blossom because things are damaged and vulnerable—not in a neoliberal logic of growth—but in the modes of attending to, feeling for, or being affected by. Practices of maintaining, repairing, recycling, and caring point to what happens in between the cracks of modernist technologies. In fact, technologies are not simply thrown out and bought new, they also do not disappear when broken; we must learn to deal better with fragile things and the practices that ensue (Jackson, 2023).

Paying attention to and responding to what is fragile is a delicate art: Ways in which actors notice by testing or touching screens and systems; how they collaborate to find compromise; how they stick their heads together to test the future. Fragility needs touch and long-term attachment, it evades being concluded or saturated, and it can deal with the fact of not knowing.

Fragility differs from similar terms like "vulnerability" as it emphasizes the local and specific rather than the general. It proposes ethics of care rather than moral judging and blaming. Attending to something vulnerable has paternalist connotations while attending to what is fragile evokes ideas about treating something carefully, being affected by it, staying attentive. Engaging in fragility means getting in touch, becoming-with rather than distancing.

Social theorist William E. Connolly writes in *The Fragility of Things*: "We are today at least as closely implicated in several nonhuman force fields […] In a world more scientifically and technically advanced we are not that much better equipped culturally, philosophically, politically or spiritually to address these entanglements" (Connolly, 2013, 7). Following him, the fragility of things requires new ways of thinking and addressing entanglements. I find that thinking with fragility as a concept allows to do just that: An emphasis on the precarity of conviviality and relation; without jumping to normative blame (how things could become better) or rushing to solutions.

Within feminist studies, Lisa Diedrich (2005) suggests attending to fragility and failure in medical narrations of cancer. She proposes that while failure and fragility have a place in Western philosophy and thinking—mostly in big stories of ascendance—particularly medical narratives convey heroic individuals' stories of overcoming and controlling diseases through medical intervention. "The steady upward curve is the favoured contour", in contemporary Western culture, she writes leaning on Jackie Stacey (Diedrich, 2005, 144). It is the crucial objective of medical narration that failure and illness must be left behind quickly. And by doing so, these medical narratives tend to oversee and discourage stories of suffrage, continued chaos, pain, or failure (Diedrich, 2005; Stacey, 1997; Rose, 1997). Reporting on her condition, Stacey tells personal stories after her chemotherapy treatment has officially concluded but her life changed drastically with a still uncertain outcome. Medical narratives might include such stories as socio-psychological factors as long as they risk patient's improvement, but if not, they simply vanish from medical narratives (Diedrich, 2005).

It is no surprise that both Stacey and Rose, that Diedrich draws on, speak of bodily fluids, feces, vomit, and other "monsters" that are usually not part of heroic stories of overcoming a disease, but the harsh realities of fragile, chaotic, and suffering bodies (Diedrich, 2005, 145). It is this shadowy side, the unknown, messy, and ugly side of modernity that we tend to overshadow with stories of progress, improvement, and success. Stacey, Diedrich writes, proposes an "art of being ill"; the art of survival when things are damaged. And it requires art: both practical and hands-on competencies and sense for the aesthetic and good. Stacey's book ends with a sober and chilling last chapter on "Endings", which discomforts the reader with the possibility of death (Diedrich, 2005, 149).

Telling the shadowy stories of technologies however may entail additional complications. Whereas ugly stories of disease are often discouraged, they are generally well-known and accepted as suffering. In technologies, reporting of irregular technological practice is condemned and eradicated. "Bad" security compliance is judged and blamed. So, finding hope within ugly technological practices requires additional need to appreciate what will not be overcome. Attending to what is fragile includes just that: Staying hopeful while developing an austere understanding that technological practices are too undesirable, broken, frail, or susceptible. As Tsing and colleagues note, we can avoid letting the same story of progress continue to drive our imagination of the possible (Tsing et al., 2017). This does not mean resignation, nor hype. Danyi (2022) has termed it a "melancholic" practice, composing what is through the eyes of what was. Tsing (2015) wants to train our "arts of noticing" and trains sensitivities of how past and present get entwined (Tsing et al., 2017), while Haraway suggests being more attentive to modes of enduring and "staying with troubles" than overcoming them (Haraway, 2016). Training to stay, to notice, to be melancholic is necessary for digital security. It allows seeing the craft of hesitation, ambiguity, and imperfection.

Literature that has attended to such tactics has provided so far valuable examples in environmental anthropology and nature-culture relation. Introducing them to high-tech and critical infrastructures allows widening the radical relations approach that such literature provides. Jackson (2014) has introduced the notion of "broken-world-thinking" to emphasize the practices of repair and maintenance rather than on those of innovation and renewal. He urges us to exercise such thinking by focusing on the uncertainties, difficulties, fragmentation, and decay present in contemporary information technology than on the novelty, growth, and progress. He contends that such a way of thinking is both normative and ontological as well as empirical and methodological. I want to take up his urge in this exact way: as both a normative and ontological provocation and an empirical and methodological one. I argue that attending to digital security is such a practice of attending to the broken rather than the new, a practice that is fundamentally open-ended and with no clear end in sight. Attending to digital security empirically as I did in the previous chapter hence provokes our ontological and normative understanding of information technology. In a world of shiny technological systems, digital security reminds us of the rusting, decaying, and constant verge of collapsing.

Some analogies can be identified. In Cohn's (2016) analysis of a high-tech satellite that faces its eternal shutdown, she argues to emphasize the satellite's earthly relations with developers and scientists. To Cohn, it is worth exploring the decay of high-tech machines as modes of action shift from creation and progress to mitigation and finalizing. This practice is not without hope, it is not fatalistic. With the satellite, some of the scientists and engineers also retired, putting their careers to rest. This makes the end of the technology also the end of the need for specific expertise and knowledge. What will happen to the satellite after it is shut down is also part of the preparing work, as it will end as space waste floating around the orbit, hopes for recycling and seeing the satellite in orbit fosters ways of continuing after the shutdown.

The chapters have presented practices of testing, tinkering, training, and performing in innovation processes. Through the description of these practices what is proposed is not one coherent, comprehensive, or representative picture of digital security, but what can be taken is a mode of theorizing digital security differently: Taking different lenses in each chapter that stand in friction with each other or are being inconsistent when looking back allows a new mode of thinking about digital security.

In "Testing", digital security was enacted as various patterns of relating, here viewed through the scope of testing practices. It was argued that it needs vocabulary to capture the work of determined security officers than bridge local adaption and mandated, formal security processes. More than simply being fluid, in the vocabulary of topologies, digital security was fragile, requiring the tedious work of security officers to actively make local practice accountable to formal procedure. But the same officers were considered blaming and shaming in "Tinkering" where two developers subverted formal regulation to a degree to care for the new technology and their colleagues. In this episode, theorizing digital security was concerned with pointing to the shifting responsibility between the two, about staying in touch with but also diverging from formal rules. Do the two fit together? In a way, yes, because they both offer a perspective on bending (yet not breaking) the rules, but they have done so with different scopes.

The inconsistency gets bigger between "Testing", "Tinkering", and "Training". Whereas the first two have both attended to practices of preventing security incidents and have done so "as good as possible" being aware of the precarity of their situation; "Training" is concerned with preparing. A complete shift of perspectives where digital security is no longer eager to prevent an incident, but rather eager to be prepared as if the other two practices had already failed. Wrangling our heads through practices of such arbitrary logics is part of fragility as a conceptual tool: it allows us to make visible tensions and ambiguity.

The book has engaged in theorizing rather than providing a theory of fragile computing. Theorizing meant taking a new conceptual lens in every chapter, letting loose ends of the previous chapters live on uneasily. This is for the writing of this book. But theories had another agency. I used them in the field to explain what my perception was of a setting or mirror to interlocutors what my interest/position was. My interlocutors were often interested in my opinion on a matter. As mentioned in the introductory paragraphs, they saw me as an ally (and later as a spy for the security department), someone to translate between digital security and other departments. Here, theories became my concrete contribution, my repertoire that I was able to contribute. The notion of "practices" and "achieving" security was copied by my interlocutors. They used them to describe what "actually" happens and that digital security "is not simply there, but needs to be done", pointing to an astonishing reworking of my concepts.

I have proposed to think with four tested STS themes: topologies, care, preparedness, and performativity in connection with digital security. All invite us to think of fragile computing not as a mere output of correct technical configuration and user compliance, but as various enactments of shifting relations; and of non-binary configurations of in/secure, non/access, ir/responsibility, etc. Fragile computing does not mean "bad" computing; it means engaging with what is good and bad awkwardly and in perpetual attachment. Rather than explaining fragile computing once and for all, or providing a coherent theory, the chapters in this book permit us to think about digital security as a practical process in which technologies are created and stabilized, and become fragile. The chapters offer stories of becoming rather than being (secure, insecure, fragile). Crucially, neither enactment is more "factual" or "better" than another—this misses the very idea—but rather provides specific theoretical repertoires to understand the doing and undoing of digital security in critical infrastructures like Powino and Covy. Neither are the four chapters coherent or encompassing in any way; others are probable, and I am urging to further multiply them.

Although, closure is not the aim of this chapter, I want to suggest four pragmatic tactics that can be followed to engage in continuous debates about digital security as fragile computing:[1]

- **Tactic 1:** Trace the socio-material networks that bring digital security into being. Ask how-questions: How does something become in/secure? How do actors perform securing tasks?
- **Tactic 2:** Go beyond asking for digital security in the form of "problems" and "solutions" in order to look at concrete practices in which it is encountered. Not every practice reacts to a problem; not every problem asks for a solution. It is much more common that daily accidents and insecurities are normal and mitigated through a series of unanticipated practices. Consult your local ethnographers to allow interlocutors a voice in defining situated questions and problems. Do not simply ask ethnographers to "bring" them to you, but discuss them, open up, and critique each other's vocabulary.
- **Tactic 3:** When creating any security measure, ask for whom, for what, where, and when. Questions like these allow locating security in a specific socio-material setting. However, even if those questions are answered and technologies build with much local sensitivity, never expect a solution will ever just "work". All technologies, including security mechanisms, are subject to situated adaptations, will become part of fragile arrangements.
- **Tactic 4:** This is related to Tactic 2 but probes a different inquiry: Instead of rushing to solutions, ask and explore if solving an issue is desirable and find out more about how it is dealt with. Many practices dwell in imperfection and in-between solutions, something that must be taken seriously, and supported if possible. Tolerate ambiguity.

Recent literature in anthropology has made it possible to capture the co-existence of progress and innovation with alternative modes of action

[1] Formulating a set of tactics to convey specific nuances of one's work is not a new tactic in STS or STS; I lean here on fabulous work by Philip et al. (2012).

that are tailored toward hesitation, care, situated obligation, and noticing. Investigating fragile computing combines STS and anthropology's growing interest in understanding how worlds are "holding" together while being under immense pressure and in a consistent possibility of failure, causing moral and spatial logics to collapse (Gan & Tsing, 2018). Fragile computing offers a counter-narrative to success stories of technologies. Most notably, that technological practice involves thinking with the possibility of failure, of ending, of decay—and the alternative moral imperatives attached to this possibility. As I have indicated in the chapters of this book, digital security involves acting in uncertainty, waiting, and preparing, often not with a clear light on the horizon. Fragile computing points to the alternative side of technologies such as the *Data Lake* that are considered "new" and "innovative", but at the same time are frail, broken, and fragile from the beginning.

I have not practiced a fatalistic tone in this book. Rather, while dwelling in insecurity, I have portrayed fragile computing as a productive, creative, and cherished practice. Making compromises, keeping open, caring, preparing. We may develop a hopeful pessimism in matters of digital security, one that will help us appreciate the fragility of our technological lives.

References

Anand, N., Gupta, A., & Appel, H. (Eds.). (2018). *The promise of infrastructure.* Duke University Press.

Appadurai, A., & Alexander, N. (2020). *Failure.* Polity Press.

Clark, A., & Haraway, D. (2018). *Making kin not population - Reconceiving generations.* Prickly Paradigm Press.

Cohn, M. L. (2016). Convivial decay: Entangled lifetimes in a geriatric infrastructure. In *CSCW'16 Proceedings and companion of the ACM conference on computer-supported cooperative work and social computing,* San Francisco, CA, pp. 1511–1523.

Connolly, W. E. (2013). *The fragility of things: Self-organizing processes, neoliberal fantasies, and democratic activism.* Duke University Press. https://doi.org/10.2307/j.ctv123x75n

Danyi, E. (2022). *Melancholy democracy: Politics beyond hope and despair.* Habilitation submitted, Institut für Soziologie, Goethe-Universität.
De Laet, M., Driessen, A., & Vogel, E. (2021). Thinking with attachments: Appreciating a generative analytic. *Social Studies of Science, 51*(6), 799–819. https://doi.org/10.1177/03063127211048804
Denis, J., & Pontille, D. (2011). *Materiality, maintenance and fragility: The care of things*. SSRN. https://ssrn.com/abstract=1947255.
Denis, J. & Pontille, D. (2013). *Material ordering and the care of things*, CSI Working Papers Series 034, Centre de Sociologie de l'Innovation (CSI), Mines ParisTech. https://ideas.repec.org/p/emn/wpaper/034.html.
Denis, J., & Pontille, D. (2020). Maintenance et attention à la fragilité. *SociologieS*. https://doi.org/10.4000/sociologies.13936
Diedrich, L. (2005). A bioethics of failure: Antiheroic cancer narratives. In M. R. Shildrick (Ed.), *Ethics of the body: Postconventional challenges* (pp. 135–150). MIT Press.
Gan, E., & Tsing, A. (2018). How things hold. *Social Analysis, 62*(4), 102–145. https://doi.org/10.3167/sa.2018.620406
Haraway, D. J. (2016). *Staying with the trouble.* Duke University Press.
Jackson, S. J. (2014). 11 Rethinking repair. In Gillespie, T., Boczkowksi, P. & Foot, K. A. (Eds.) *Media technologies: Essays on communication, materiality, and society* (pp. 221–239). https://doi.org/10.7551/mitpress/9042.003.0015.
Jackson, S. J. (2023). Ordinary hope. In M. Taccheti, D. Papadopoulos, & M. Puig de la Bellacasa (Eds.), *Ecological reparation: Repair, remediation and resurgence in social and environmental conflict* (pp. 417–433). Bristol University Press.
Liebetrau, T., & Christensen, K. K. (2021). The ontological politics of cyber security: Emerging agencies, actors, sites and spaces. *European Journal of International Security, 6*(1), 25–43. https://doi.org/10.1017/eis.2020.10
Mol, A. (2002). *The body multiple: Ontology in medical practice*. Duke University Press. https://doi.org/10.2307/j.ctv1220nc1
Mol, A. (2006). Proving or improving: On health care research as a form of self-reflection. *Qualitative health research, 16*(3), 405–414. https://doi.org/10.1177/1049732305285856
Philip, K., Irani, L., & Dourish, P. (2012). Postcolonial computing: A tactical survey. *Science, Technology, and Human Values, 37*(1), 3–29. https://doi.org/10.1177/0162243910389594
Rose, G. (1997). *Love's work*. Vintage.
Stacey, J. (1997). *Teratologies*. Routledge.

Stengers, I. (2017). *Another science is possible* (S. Muecke, Trans.). Polity Press.
Tsing, A. L. (2004). *Friction*. Princeton University Press.
Tsing, A. L. (2015). *The mushroom at the end of the world: On the possibility of life in capitalist ruins*. Princeton University Press.
Tsing, A. L., Swanson, H., Gan, E., & Bubandt, N. (2017). *Arts of living on a damaged planet: Ghosts of the anthropocene*. University of Minnesota Press.

References

Adams, A., & Sasse, M. A. (1999). Users are not the enemy. *Communications of the ACM, 42*(12), 40–46. https://doi.org/10.1145/322796.322806

Adams, V., Murphy, M., & Clarke, A. E. (2009). Anticipation: Technoscience, life, affect, temporality. *Subjectivity, 28*(1), 246–265. https://doi.org/10.1057/sub.2009.18

Adey, P., & Anderson, B. (2011). Event and anticipation: UK civil contingencies and the space—Times of decision. *Environment and Planning A, 43*(12), 2878–2899. https://doi.org/10.1068/a43576

Adey, P., & Anderson, B. (2012). Anticipating emergencies: Technologies of preparedness and the matter of security. *Security Dialogue, 43*(2), 99–117. https://doi.org/10.1177/0967010612438432

Akrich, M. (1992). The de-scription of technical objects. In W. E. Bijker und J. Law (Hg.), *Shaping technology. Studies in sociotechnical change*, (pp. 205–224). MIT Press (Inside Technology).

Al-Matari, O. M., Helal, I. M. A., Mazen, S. A., & Elhennawy, S. (2018). Cybersecurity tools for IS auditing. In *2018 Sixth international conference on enterprise systems (ES)* (pp. 1–6). IEEE.

Anand, N., Gupta, A., & Appel, H. (Eds.). (2018). *The promise of infrastructure.* Duke University Press.

Anderson, C. (2008). The end of theory: The data deluge makes the scientific method obsolete. *Wired Magazine, 16*(7), 16–07.

L. Kocksch, *Fragile Computing*, https://doi.org/10.1007/978-981-99-9807-4

Anderson, B. (2010). Preemption, precaution, preparedness: Anticipatory action and future geographies. *Progress in Human Geography, 34*(6), 777–798. https://doi.org/10.1177/0309132510362600

Appadurai, A., & Alexander, N. (2020). *Failure*. Polity Press.

Aradau, C. (2010). Security that matters: Critical infrastructure and objects of protection. *Security Dialogue, 41*(5), 491–514. https://doi.org/10.1177/0967010610382687

Bada, M., Sasse, A. M., & Nurse, J. R. C. (2019). *Cyber Security Awareness Campaigns: Why do they fail to change behaviour?* http://arxiv.org/pdf/1901.02672v1

Baker, N. D. (2014). Role of explicit and implicit practices in the production of situated preparedness for disasters. *Natural Hazards Review, 15*(4), 5014002. https://doi.org/10.1061/(ASCE)NH.1527-6996.0000127

Baker, N. D., & Ludwig, L. G. (2018). Disaster preparedness as social control. *Critical Policy Studies, 12*(1), 24–43. https://doi.org/10.1080/19460171.2016.1214075

Balzacq, T., & Cavelty, M. D. (2016). A theory of actor-network for cybersecurity. *European Journal of International Security, 1*(2), 176–198. https://doi.org/10.1017/eis.2016.8

Barry, A., & Born, G. (2013). Interdisciplinarity: Reconfigurations of the social and natural sciences. In A. Barry & G. Born (Eds.), *Interdisciplinarity: Reconfigurations of the social and natural sciences* (pp. 1–56). Routledge.

Beck, U. (1992). *Risk society*. Sage.

Berg, M., & Timmermans, S. (2000). Orders and their others: On the constitution of universalities in medical work. *Configurations, 8*, 31–61. https://doi.org/10.1353/con.2000.0001

Bittner, E. & Garfinkel, H. (1967). "Good" organizational reasons for "bad" clinical records. In *Studies in ethnomethodology*. Prentice Hall.

Boas, F. (1919). Scientists as spies. *The Nation, 27*. https://doi.org/10.1111/j.0268-540X.2005.00359.x

Bowker, G. C. (2008). *Memory practices in the sciences*. MIT Press.

Bowker, G. C., & Star, S. L. (2000). Sorting things out: Classification and its consequences. https://doi.org/10.7551/mitpress/6352.001.0001

Bravo-Lillo, C., Cranor, L. F., Downs, J., & Komanduri, S. (2011). Bridging the gap in computer security warnings: A mental model approach. *IEEE Security and Privacy, 9*(2), 18–26. https://doi.org/10.1109/MSP.2010.198

Burdon, M., & Coles-Kemp, L. (2019). The significance of securing as a critical component of information security: An Australian narrative. *Computers & Security, 87*, S. 101601. https://doi.org/10.1016/j.cose.2019.101601

Burton, M. A. (1994). The Irrigation Management Game: A role playing exercise for training in irrigation management. *Irrigation and Drainage Systems, 7*(4), 305–318. https://doi.org/10.1007/BF00881557

Buzan, B., Wæver, O., & De Wilde, J. (1998). *Security: A new framework for analysis*. Lynne Rienner Publishers.

Caldwell, T. (2016). Making security awareness training work. *Computer Fraud and Security, 2016*(6), 8–14. https://doi.org/10.1016/S1361-3723(15)30046-4

Callon, M. (1984). Some elements of a sociology of translation: Domestication of the scallops and the fishermen of St Brieuc Bay. *The Sociological Review, 32*(1_suppl), 196–233.

Callon, M., & Law, J. (2005). On qualculation, agency, and otherness. *In: Environ Plan D, 23*(5), 717–733. https://doi.org/10.1068/d343t

Chernysheva, L., Dányi, E., de la Cadena, M., Farquhar, J., Joks, S., Kondo, S., Lai, L., Law, J., Lin, W., Magelhães, B., Martínez Medina, S., M'charek, A., Mol, A., Sezneva, O., Swanson, H., & Østmo, L. (2020). Translating a title: On other terms. *The Sociological Review, 68*(2), 283–286. https://doi.org/10.1177/0038026120905470

Clark, A., & Haraway, D. (2018). *Making kin not population - Reconceiving generations*. Prickly Paradigm Press.

Cohn, M. L. (2016). Convivial decay: Entangled lifetimes in a geriatric infrastructure. In *CSCW'16 Proceedings and companion of the ACM conference on computer-supported cooperative work and social computing*, San Francisco, CA, pp. 1511–1523.

Collier, S. J., & Lakoff, A. (2008). Distributed preparedness: The spatial logic of domestic security in the United States. *Environment and Planning D, 26*(1), 7–28. https://doi.org/10.1068/d446t

Collier, S. J., & Lakoff, A. (2015). Vital systems security: Reflexive biopolitics and the government of emergency. *Theory, Culture and Society, 32*(2), 19–51. https://doi.org/10.1177/0263276413510050

Connolly, M. (2012). Creating a campus based community emergency response team (CERT). *Community College Journal of Research and Practice, 36*(6), 448–452. https://doi.org/10.1080/10668920903018134

Connolly, W. E. (2013). *The fragility of things: Self-organizing processes, neoliberal fantasies, and democratic activism*. Duke University Press. https://doi.org/10.2307/j.ctv123x75n

Danyi, E. (2022). *Melancholy democracy: Politics beyond hope and despair*. Habilitation submitted, Institut für Soziologie, Goethe-Universität.

de Goede, M., & Randalls, S. (2009). Precaution, preemption: Arts and technologies of the actionable future. *Environment and Planning D, 27*(5), 859–878. https://doi.org/10.1068/d2608

De Laet, M., & Mol, A. (2000). The Zimbabwe bush pump. *Social Studies of Science, 30*(2), 225–263. https://doi.org/10.1177/030631200030002002

De Laet, M., Driessen, A., & Vogel, E. (2021). Thinking with attachments: Appreciating a generative analytic. *Social Studies of Science, 51*(6), 799–819. https://doi.org/10.1177/03063127211048804

Denis, J., & Pontille, D. (2011). *Materiality, maintenance and fragility: The care of things*. SSRN. https://ssrn.com/abstract=1947255.

Denis, J. & Pontille, D. (2013). *Material ordering and the care of things*, CSI Working Papers Series 034, Centre de Sociologie de l'Innovation (CSI), Mines ParisTech. https://ideas.repec.org/p/emn/wpaper/034.html.

Denis, J., & Pontille, D. (2015). Material ordering and the care of things. *Science, Technology, and Human Values, 40*(3), 338–367. https://doi.org/10.1177/0162243914553129

Denis, J., & Pontille, D. (2020). Maintenance et attention à la fragilité. *SociologieS*. https://doi.org/10.4000/sociologies.13936

Denis, J., & Pontille, D. (2023). Cultivating attention to fragility: The sensible encounters of maintenance. In *Ecological reparation* (pp. 344–361). Bristol University Press.

Deville, J. (2021). Waiting on standby: The relevance of disaster preparedness. In *Ephemera: Theory and Politics in Organization*, 21(1). https://eprints.lancs.ac.uk/id/eprint/148931/

Deville, J., & Guggenheim, M. (2018). From preparedness to risk: From the singular risk of nuclear war to the plurality of all hazards. *The British Journal of Sociology, 69*(3), 799–824. https://doi.org/10.1111/1468-4446.12291

Deville, J., Guggenheim, M., & Hrdličková, Z. (2014). Concrete governmentality: Shelters and the transformations of preparedness. *The Sociological Review, 62*(1_suppl), 183–210. https://doi.org/10.1111/1467-954X.12129

Dewey, J. (1994). *Experience and nature*. 9th printing. Open Court (The Paul Carus lectures series, 1).

Diedrich, L. (2005). A bioethics of failure: Antiheroic cancer narratives. In M. R. Shildrick (Ed.), *Ethics of the body: Postconventional challenges* (pp. 135–150). MIT Press.

Donaldson, S. E., Siegel, S. G., Williams, C. K., & Aslam, A. (2015). *Enterprise cybersecurity. How to build a successful cyberdefense program against advanced threats*. Apress Springer Science+Business Media.

Dourish, P. (2004). What we talk about when we talk about context. *Personal and Ubiquitous Computing, 8*(1), 19–30. https://doi.org/10.1007/s00779-003-0253-8

Dourish, P. (2006). Implications for design. In R. Grinter (Ed.), *Proceedings of the SIGCHI conference on human factors in computing systems* (pp. 541–550). ACM.

Dourish, P. (2017). *The stuff of bits. An essay on the materialities of information.* The MIT Press. https://ieeexplore.ieee.org/book/7932719.

Dourish, P., Grinter, R. E., La Delgado Flor, J., & Joseph, M. (2004). Security in the wild: User strategies for managing security as an everyday, practical problem. *Personal and Ubiquitous Computing, 8*(6), 391–401. https://doi.org/10.1007/s00779-004-0308-5

Edwards, P. N. (2010). *A vast machine: Computer models, climate data, and the politics of global warming*. MIT Press.

Edwards, P. N. (2013). *A vast machine.* MIT Press.

Edwards, P. N., Bowker, G. C., Jackson, S. J., & Williams, R. (2009). Introduction: An Agenda for infrastructure studies. *Journal of the Association for Information Systems, 10*(5), 364–374. https://doi.org/10.17705/1jais.00200

Edwards, P., Mayernik, M., Batcheller, A., Bowker, G., & Borgman, C. (2011). Science friction: Data, metadata, and collaboration. *Social Studies of Science, 41*(5), 667–690. https://doi.org/10.1177/0306312711413314. Retrieved from https://escholarship.org/uc/item/2tj4g7sc

Emerson, R. M., Fretz, R. I., & Shaw, L. L. (2011). *Writing ethnographic field-notes* (2nd ed.). The University of Chicago Press.

Fisher, B., & Tronto, J. (1990). Toward a feminist theory of caring. In E. K. Abel & M. K. Nelson (Eds.), *Circles of care: Work and identity in women's lives* (pp. 35–62). State University of New York Press.

Forsythe, D. E. (1999). "It's just a matter of common sense": Ethnography as invisible work. *Computer Supported Cooperative Work, 8*(1–2), 127–145. https://doi.org/10.1023/A:1008692231284

Fortun, M., Fortun, K. & Marcus, G. E. (2017). Computers in/and anthropology: The poetics and politics of digitization. In Larissa H.; Horst, H. A.; Galloway, A. & Bell, G. (Eds.) *The Routledge companion to digital ethnography* (pp. 37–46). Routledge Taylor & Francis Group. https://www.taylorfrancis.com/chapters/edit/10.4324/9781315673974-11/computers-anthropology-poetics-politics-digitization-mike-fortun-kim-fortun-george-marcus

Gan, E., & Tsing, A. (2018). How things hold. *Social Analysis, 62*(4), 102–145. https://doi.org/10.3167/sa.2018.620406

Gilligan, C. (1982). New maps of development: New visions of maturity. *The American Journal of Orthopsychiatry, 52*(2), 199–212. https://doi.org/10.1111/j.1939-0025.1982.tb02682.x

Grinberg, Y. (2016). *Is data singular or plural?* CASTAC. http://blog.castac.org/2016/10/is-data/.

Hammersley, M. (1990). What's wrong with ethnography? The myth of theoretical description. *Sociology, 24*(4), 597–615. https://doi.org/10.1177/0038038590024004003

Hanseth, O., Monteiro, E., & Hatling, M. (1996). Developing information infrastructure: The tension between standardization and flexibility. *Science, Technology, & Human Values, 21*(4), 407–426. https://doi.org/10.1177/016224399602100402

Haraway, D. (1988). Situated knowledges: The science question in feminism and the privilege of partial perspective. *Feminist Studies, 14*(3), 575–599. https://doi.org/10.2307/3178066

Haraway, D. J. (2016). *Staying with the trouble.* Duke University Press.

Hedström, K., Dhillon, G., & Karlsson, F. (2010). Using actor network theory to understand information security management. In K. Rannenberg (Ed.), *Security and privacy - Silver linings in the cloud: 25th IFIP TC-11 international information security conference, SEC 2010,* Held as Part of WCC 2010, Brisbane, September 20–23, 2010 (pp. 43–54).

Heidenstrøm, N. (2019). Informal household preparedness: methodological approaches to everyday practices. *Journal of Risk Research, 23*(3), 379–397. https://doi.org/10.1080/13669877.2019.1569106

Heuts, F., & Mol, A. (2013). What is a good tomato? A case of valuing in practice. *Valuation Studies, 1*(2), 125–146. https://doi.org/10.3384/vs.2001-5992.1312125

Hirschauer, S. (2001). Ethnografisches Schreiben und die Schweigsamkeit des Sozialen/Ethnographic Writing and the Silence of the Social. *Zeitschrift für Soziologie, 30*(6), 429–451. https://doi.org/10.1515/zfsoz-2001-0602

Houston, L., Jackson, S. J., Rosner, D. K., Ahmed, S. I., Young, M., & Kang, L. (2016). Values in repair. In J. Kaye, A. Druin, C. Lampe, D. Morris, & J. P. Hourcade (Eds.), *#chi4good: CHI 2016,* (pp. 1403–1414). San Jose, CA, May 7–12.

Hutchins, E. (1996). *Cognition in the wild.* Bradford Books.

Jack, M., & Jackson, S. J. (2016). Logistics as care and control. In J. Kaye, A. Druin, C. Lampe, D. Morris, & J. P. Hourcade (Eds.), *#chi4good: CHI 2016.* San Jose, CA, May 7–12 (pp. 2209–2219). Association for Computing Machinery Inc. (ACM).

Jackson, S. J. (2014). 11 Rethinking repair. In Gillespie, T., Boczkowksi, P. & Foot, K. A. (Eds.) *Media technologies: Essays on communication, materiality, and society* (pp. 221–239). https://doi.org/10.7551/mitpress/9042.003.0015

Jackson, S. J. (2023). Ordinary hope. In M. Taccheti, D. Papadopoulos, & M. Puig de la Bellacasa (Eds.), *Ecological reparation: Repair, remediation and resurgence in social and environmental conflict* (pp. 417–433). Bristol University Press.

Kemmer, L., Kühn, A., Otto, B., & Weber, V. (2021). Standby: Organizing modes of in|activity. *Ephemera, 21*(1), 1–20.

Killcrece, G., Kossakowski, K. P., & Ruefle, R. (2003). *Organizational models for computer security incident response teams (CSIRTs).* Carnegie-Mellon University, Software Engineering Institute. Retrieved from https://resources.sei.cmu.edu/asset_files/Handbook/2003_002_001_14099.pdf.

Kitchin, R. (2014). *The data revolution: Big data, open data, data infrastructures and their consequences.* Sage. https://doi.org/10.4135/9781473909472

Klimburg-Witjes, N., & Wentland, A. (2021). Hacking humans? Social engineering and the construction of the "deficient user" in cybersecurity discourses. *Science, Technology, and Human Values, 46*(6), 1316–1339. https://doi.org/10.1177/0162243921992844

Knox, H. (2020). *Thinking like a climate.* Duke University Press.

Knox, H., & Nafus, D. (Eds.). (2018). *Ethnography for a data-saturated world.* Manchester University Press.

Kocksch, L. (2020). Knowing IT-security emergencies. Die Herstellung von Unsicherheit. In: *Astrid Wiedmann/Katherin Wagenknecht/Philipp Goll/ Andreas Wagenknecht, Wie forschen mit den »Science and Technology Studies«?* (pp. 175–210). Transcript Verlag. https://doi.org/10.14361/9783839443798-007

Kocksch, L., Korn, M., Poller, A., & Wagenknecht, S. (2018). Caring for IT Security: Accountabilities, moralities, and oscillations in IT security practices. *Proceedings of the ACM on Human-Computer Interaction, 2*(CSCW), Article 92. https://doi.org/10.1145/3274361

Kreitz, M. (2020). Security by design in software engineering. *SIGSOFT Software Engineering Notes, 44*(3), 23. https://doi.org/10.1145/3356773.3356798

Lakoff, A. (2005). From disaster to catastrophe: The limits of preparedness. In *Online forum and essays.* Social Science Research Council.

Lakoff, A. (2008). The generic biothreat, or, how we became unprepared. *Cultural Anthropology, 23*(3), 399–428

Lampland, M. (Ed.). (2009). *Standards and their stories. How quantifying classifying and formalizing practices shape everyday life*. Cornell University Press.
Latimer, J. (2000). Socialising disease: Medical categories and inclusion of the aged. *The Sociological Review, 48*(3), 383–407. https://doi.org/10.1111/1467-954X.00222
Latour, B. (1990). Technology is society made durable. *The Sociological Review, 38*(S1), 103–131.
Latour, B. (1992). Where are the missing masses? The sociology of a few mundane artifacts. In W. Bijker & J. Law (Eds.), *Shaping technology-building society: Studies in sociotechnical change* (pp. 225–259). MIT Press.
Latour, B. (1996). *Aramis or the love of technology*. Harvard University Press.
Latour, B. (1999). *Pandora's hope. Essays on the reality of science studies*. Harvard University Press.
Latour, B., & Venn, C. (2002). Morality and technology. *Theory, Culture and Society, 19*(5–6), 247–260. https://doi.org/10.1177/026327602761899246
Latour, B., Jensen, P., Venturini, T., Grauwin, S., & Boullier, D. (2014). The whole is always smaller than its parts'. *A Digital Test of Gabriel Tarde's Monads*, 31.07.2014. https://hal-sciencespo.archives-ouvertes.fr/hal-01053567/document.
Law, J. (1987). On the social explanation of technical change: The case of the Portuguese maritime expansion. *Technology and Culture, 28*(2), S. 227. https://doi.org/10.2307/3105566
Law, J. (1999). After ant: Complexity, naming and topology. *The Sociological Review, 47*(1_suppl), 1–14. https://doi.org/10.1111/j.1467-954X.1999.tb03479.x
Law, J. (2000). *Ladbroke Grove, or how to think about failing systems*. http://www.comp.lancs.ac.uk/sociology/papers/Law-Ladbroke-Grove-Failing-Systems.pdf.
Law, J. (2002). *Aircraft stories. Decentering the object in technoscience*. Duke University Press (Science and cultural theory). http://site.ebrary.com/lib/alltitles/docDetail.action?docID=10207714.
Law, J., & Lin, W. (2010). Cultivating disconcertment. *The Sociological Review, 58*(2_suppl), 135–153. https://doi.org/10.1111/j.1467-954X.2011.01966.x
Law, J., & Mol, A. (2001). Situating technoscience: An inquiry into spatialities. *Environment and Planning D: Society and Space, 19*(5), 609–621. https://doi.org/10.1068/d243t

Law, J., & Mol, A. (2002a). Local entanglements or utopian moves: An inquiry into train accidents. *The Sociological Review, 50*(1_suppl), 82–105. https://doi.org/10.1111/j.1467-954x.2002.tb03580.x

Law, J., & Mol, A. (Eds.). (2002b). *Complexities*. Duke University Press.

Law, J., & Mol, A. (2020). Words to think with: An introduction. *The Sociological Review, 68*(2), 263–282. https://doi.org/10.1177/0038026120905452

Law, J., & Singleton, V. (2005). Object lessons. *Organization, 12*(3), 331–355. https://doi.org/10.1177/1350508405051270

Leonelli, S., Rappert, B., & Davies, G. (2016). Data shadows. *In Science, Technology, and Human Values, 42*(2), 191–202. https://doi.org/10.1177/0162243916687039

Liburkina, R. (2021). Extraordinary ethnographic encounters in extraordinary times: A plea for experimental interventions in more-than-business relations. *Kulturanthropologie Notizen, 83*, 14–26.

Liebetrau, T., & Christensen, K. K. (2021). The ontological politics of cyber security: Emerging agencies, actors, sites and spaces. *European Journal of International Security, 6*(1), 25–43. https://doi.org/10.1017/eis.2020.10

Lindell, M. K., Arlikatti, S., & Prater, C. S. (2009). Why people do what they do to protect against earthquake risk: Perceptions of hazard adjustment attributes. *Risk Analysis, 29*(8), 1072–1088. https://doi.org/10.1111/j.1539-6924.2009.01243.x

Loukissas, Y. A. (2019). *All data are local*. MIT Press.

Marres, N. (2012). *Material participation. Technology the environment and everyday publics*. Palgrave Macmillan.

Martin, A., Myers, N., & Viseu, A. (2015). The politics of care in technoscience. *Social studies of science, 45*(5), 625–641. https://doi.org/10.1177/0306312715602073

McCrohan, K. F., Engel, K., & Harvey, J. W. (2010). Influence of awareness and training on cyber security. *Journal of Internet Commerce, 9*(1), 23–41. https://doi.org/10.1080/15332861.2010.487415

Mol, A. (2003). *The body multiple: Ontology in medical practice*. Duke University Press. https://doi.org/10.1215/9780822384151

Mol, A. (2003). *The body multiple*. Duke University Press.

Mol, A. (2006). Proving or improving: On health care research as a form of self-reflection. *Qualitative health research, 16*(3), 405–414. https://doi.org/10.1177/1049732305285856

Mol, A. (2008). *The logic of care*. Routledge.

Mol, A., & Law, J. (1994). Regions, networks and fluids: Anaemia and social topology. *Social Studies of Science, 24*(4), 641–671. https://doi.org/10.1177/030631279402400402

Mol, A., Moser, I., & Pols, J. (2010). Care: Putting practice into theory. In *Care in practice* (pp. 7–26). https://doi.org/10.14361/transcript.9783839414477.7

Moser, I., & Law, J. (2006). Fluids or flows? Information and qualculation in medical practice. *Info Technology and People, 19*(1), 55–73. https://doi.org/10.1108/09593840610649961

Muniesa, F., & Linhardt, D. (2011). Trials of explicitness in the implementation of public management reform. *Critical Perspectives on Accounting, 22*(6), 550–566. https://doi.org/10.1016/j.cpa.2011.06.003

National Institute of Standards and Technology (NIST). (2018). *Security considerations in the system development life cycle*. NIST Special Publication 800-64 Revision 2. October 2008. Retrieved from https://nvlpubs.nist.gov/nistpubs/legacy/sp/nistspecialpublication800-64r2.pdf

Nieves Delgado, A., & Kocksch, L. (2022). Privacy in social media. *Techné: Research in Philosophy and Technology, 26*(2), 257–284.

Niewöhner, J. (2016). Co-laborative anthropology: Crafting reflexivities experimentally. *Etnologinen tulkinta ja analyysi*, 81–124. (in Finnish).

Nowviskie, B. (2015). Digital humanities in the anthropocene. *Digital scholarship in the humanities, 30*(suppl_1), i4–i15.

Palombo, H., Ziaie Tabari, A., Lende, D., Ligatti, J., & Ou, X. (2020). An ethnographic understanding of software (in) security and a co-creation model to improve secure software development. In *Proceedings of the sixteenth USENIX conference on usable privacy and security (SOUPS'20)* (pp. 205–220). USENIX Association.

Pandian, A. (2019). *A possible anthropology: Methods for uneasy times*. Duke University Press.

Passi, S., & Jackson, S. J. (2018). Trust in data science: Collaboration, translation, and accountability in corporate data science projects. In *Proceedings of the ACM on human-computer interaction*, 2(CSCW), 136. https://doi.org/10.1145/3274405

Perrow, C. (1999). *Normal accidents*. Princeton University Press.

Philip, K., Irani, L., & Dourish, P. (2012). Postcolonial computing: A tactical survey. *Science, Technology, and Human Values, 37*(1), 3–29. https://doi.org/10.1177/0162243910389594

Pinel, C., Prainsack, B., & McKevitt, C. (2020). Caring for data. Value creation in a data-intensive research laboratory. *Social Studies of Science, 50*(2), 175–197. https://doi.org/10.1177/0306312720906567

Pipek, V., & Wulf, V. (2009). Infrastructuring: Toward an integrated perspective on the design and use of information technology. *Journal of the Association for Information Systems, 10*(5), 447–473. https://doi.org/10.17705/1jais.00195

Poller, A., Kocksch, L., Türpe, S., Epp, F. A., & Kinder-Kurlanda, K. (2017). Can security become a routine? A study of organizational change in an agile software development group. In *Proceedings of the 2017 ACM conference on computer supported cooperative work and social computing (CSCW '17)*, 2489–2503. Association for Computing Machinery. https://doi.org/10.1145/2998181.2998191

Puig de la Bellacasa, M. (2011). Matters of care in technoscience: Assembling neglected things. *Social Studies of Science, 41*(1), 85–106.

Puig de La Bellacasa, M. (2012). 'Nothing comes without its world': Thinking with care. *The Sociological Review, 60*(2), 197–216. https://doi.org/10.1111/j.1467-954X.2012.02070.x

Puig de La Bellacasa, M. (2017). *Matters of care: Speculative ethics in more than human worlds*. University of Minnesota Press. (Posthumanities, v.41).

Rahim, N. H. A., Hamid, S., Mat Kiah, M. L., Shamshirband, S., & Furnell, S. (2015). A systematic review of approaches to assessing cybersecurity awareness. *Kybernetes, 44*(4), 606–622. https://doi.org/10.1108/K-12-2014-0283

Ramakrishnan, K., O'Reilly, K., & Budds, J. (2021). The temporal fragility of infrastructure: Theorizing decay, maintenance, and repair. *Environment and Planning E: Nature and Space, 4*(3), 674–695. https://doi.org/10.1177/2514848620979712

Reckwitz, A. (2007). Practice theory. In G. Ritzer (Ed.) *The Blackwell encyclopedia of sociology*. https://doi.org/10.1002/9781405165518.wbeosp125

Rose, G. (1997). *Love's Work*. Vintage.

Sabillon, R., Serra-Ruiz, J., Cavaller, V., & Cano, J. (2017). A comprehensive cybersecurity audit model to improve cybersecurity assurance: The CyberSecurity Audit Model (CSAM). In *Proceedings of the 2017 international conference on information systems and computer science (INCISCOS 2017)* (pp. 253–259). IEEE.

Schatzki, T. R., Knorr-Cetina, K., & von Savigny, E. (2000). *The practice turn in contemporary theory*. Routledge.

Shostack, A. (2014). *Threat modeling. Designing for security*. Wileys. https://learning.oreilly.com/library/view/-/9781118810057/?ar

Singleton, V. (2010). Good farming control or care? In A. Mol, I. Moser, & J. Pols (Eds.), *Care in practice. On tinkering in clinics, homes and farms* (pp. 235–256). Transcript.

Slayton, R., & Clarke, B. (2020). Trusting infrastructure: The emergence of computer security incident response, 1989–2005. *Technology and Culture, 61*(1), 173–206. https://doi.org/10.1353/tech.2020.0036

Sørensen, E. (2009). *The materiality of learning: Technology and knowledge in educational practice (Learning in doing: Social, cognitive and computational perspectives).* Cambridge University Press. https://doi.org/10.1017/CBO9780511576362

Sørensen, E., & Kocksch, L. (2021). Data durabilities: Towards conceptualizations of scientific long-term data storage. *Engaging Science, Technology, and Society, 7*(1), 12–21. https://doi.org/10.17351/ests2021.777

Spencer, M., & Pizio, D. (2023). The de-perimeterisation of information security: The Jericho Forum, zero trust, and narrativity. *Social Studies of Science*, 3063127231221107. Advance online publication. https://doi.org/10.1177/03063127231221107

Squires, S., & Shade, M. (2015). People, the weak link in cyber-security: Can ethnography bridge the gap? *Ethnographic Praxis in Industry Conference Proceedings, 1*, 47–57. https://doi.org/10.1111/1559-8918.2015.01039

Stacey, J. (1997). *Teratologies*. Routledge.

Stachel, R. D., & DeLaHaye, M. (2015). Security breaches in healthcare data: An application of the actor-network theory. *Issues in Information Systems, 16*(II), 185–194.

Star, S. L. (1990). Power, technology and the phenomenology of conventions: On being allergic to onions. *The Sociological Review, 38*(1_suppl), 26–56. https://doi.org/10.1111/j.1467-954X.1990.tb03347.x

Star, S. L. (1999). The ethnography of infrastructure. *American Behavioral Scientist, 43*(3), 377–391. https://doi.org/10.1177/00027649921955326

Star, S. L., & Griesemer, J. R. (1989). Institutional ecology, 'translations' and boundary objects: Amateurs and professionals in Berkeley's museum of vertebrate zoology, 1907–39. *Social Studies of Science, 19*(3), 387–420. https://doi.org/10.1177/030631289019003001

Star, S. L., & Ruhleder, K. (1994). Steps towards an ecology of infrastructure. In J. B. Smith, F. D. Smith, & T. W. Malone (Eds.), *Proceedings of the 1994 ACM conference on computer supported cooperative work* (pp. 253–264). ACM.

Star, S. L., & Strauss, A. (1999). Layers of silence, arenas of voice: The ecology of visible and invisible work. *Computer Supported Cooperative Work (CSCW), 8*, 9–30. https://doi.org/10.1023/A:1008651105359

Stengers, I. (2017). *Another science is possible* (S. Muecke, Trans.). Polity Press.

Strathern, M. (1991). *Partial connections.* Rowman & Littlefield (ASAO special publications, 3).

Strathern, M. (2000). The tyranny of transparency. *British Educational Research Journal, 26*(3), 309–321. https://doi.org/10.1080/713651562

Strathern, M. (2002). Abstraction and decontextualization. An anthropological comment. In S. Woolgar (Ed.), *Virtual society?: Technology, cyberbole, reality* (pp. 302–314).

Suchman, L. (2006). Human-machine reconfigurations: Plans and situated actions. In *Learning in doing: Social, cognitive and computational perspectives* (2nd ed.). Cambridge University Press. https://doi.org/10.1017/CBO9780511808418

Sundaramurthy, S. C., Case, J., Truong, T., Zomlot, L., & Hoffmann, M. (2014). A tale of three security operation centers. In *Proceedings of the 2014 ACM workshop on security information workers (SIW' 14)* (pp. 43–50). Association for Computing Machinery. https://doi.org/10.1145/2663887.2663904

Taylor, A. R. E. (2021). Standing by for data loss: Failure, preparedness and the cloud. *Ephemera: Theory and Politics in Organization, 21*(1), 59–93.

Tempini, N. (2016). *Science through the "Golden Security Triangle"*. Information Security and Data Journeys in Data-intensive Biomedicine: Association for Information Systems (AIS). https://ore.exeter.ac.uk/repository/bitstream/10871/24795/1/InfoSec%26Research_ICIS_2016_Final_NT.pdf.

Tempini, N., & Leonelli, S. (2018). Concealment and discovery: The role of information security in biomedical data re-use. *Social Studies of Science, 48*(5), 663–690. https://doi.org/10.1177/0306312718804875

Timmermans, S., & Berg, M. (1997). Standardization in action: Achieving local universality through medical protocols. *Social Studies of Science, 27*(2), 273–305. https://doi.org/10.1177/030631297027002003

Tirumala, S. S., Sarrafzadeh, A., & Pang, P. (2016). A survey on internet usage and cybersecurity awareness in students. In *2016 14th annual conference on privacy, security and trust (PST)* (pp. 223–228). IEEE. https://doi.org/10.1109/PST.2016.7906950

Tøndel, I. A., Cruzes, D. S., & Jaatun, M. G. (2020). Using situational and narrative analysis for investigating the messiness of software security. In *Proceedings of the 14th ACM. ESEM '20: ACM/IEEE international symposium on empirical software engineering and measurement* (pp. 1–6). ACM Digital Library. https://doi.org/10.1145/3382494.3410701

Toombs, A. L., Bardzell, S., & Bardzell, J. (2015). The proper care and feeding of hackerspaces. In *Proceedings of the 33rd annual ACM conference on human factors in computing systems - CHI '15* (pp. 629–638). ACM Press. https://doi.org/10.1145/2702123.2702284

Tronto, J. C. (1998). An ethic of care. *Generations: Journal of the American Society on Aging, 22*(3), 15–20.

Tsing, A. L. (2004). *Friction*. Princeton University Press.

Tsing, A. L. (2011). *Friction: An ethnography of global connection*. Princeton University Press.

Tsing, A. L. (2015). *The mushroom at the end of the world: On the possibility of life in capitalist ruins*. Princeton University Press.

Tsing, A. L., Swanson, H., Gan, E., & Bubandt, N. (2017). *Arts of living on a damaged planet: Ghosts of the anthropocene*. University of Minnesota Press.

Verran, H. (2001). *Science and an African logic* (2nd ed.). University of Chicago Press.

Vinsel, L. (2023). *"How to be a better reactionary: Time and knowledge in technology regulation" Essay for Workshop "Politics of Controlling Powerful Technologies"*. Oct 17, 2023. https://sts-news.medium.com/how-to-be-a-better-reactionary-1630b5098fbc

Williams, P. A. H. & Coles-Kemp, L. (2019). Teaching as a collaborative practice: Reframing security practitioners as navigators. In Zhigeng Pan, A. D., Müller, C.W., Mingmin Zhang, A. E. R., & Kifayat, K. (Eds.), *Transactions on edutainment XV, Bd. 11345* (pp. 108–128). Springer (Springer eBooks Computer Science, 11345).

Winthereik, B. R. (2020). Is ANT's radical empiricism ethnographic? In A. Blok, I. Farias, & C. Roberts (Eds.), *The Routledge companion to actor-network theory* (pp. 24–33). Routledge.

Wolf, M., & Hall, K. (2018). Cyborg preparedness: Incorporating knowing and caring bodies into emergency infrastructures. *Medical Anthropology, 37*(6), 486–498. https://doi.org/10.1080/01459740.2018.1485022

Index[1]

[1] Note: Page numbers followed by 'n' refer to notes.

L. Kocksch, *Fragile Computing*, https://doi.org/10.1007/978-981-99-9807-4

C

D

SPRINGER NATURE

GPSR Compliance

The European Union's (EU) General Product Safety Regulation (GPSR) is a set of rules that requires consumer products to be safe and our obligations to ensure this.

If you have any concerns about our products, you can contact us on ProductSafety@springernature.com

In case Publisher is established outside the EU, the EU authorized representative is:

Springer Nature Customer Service Center GmbH
Europaplatz 3
69115 Heidelberg, Germany

The manufacturer's authorised representative in the EU is Springer Nature Customer Service Centre GmbH, Europaplatz 3, 69115 Heidelberg, Germany. If you have any concerns regarding our products, please contact ProductSafety@springernature.com

Printed and bound by CPI Group (UK) Ltd, Croydon, CR0 4YY
01/10/2026
02225903-0006